IGA AND KOKA NINJA SKILLS

IGA AND KOKA NINJA SKILLS

THE SECRET SHINOBI SCROLLS OF CHIKAMATSU SHIGENORI

INCLUDING A COMMENTARY ON SUN TZU'S 'USE OF SPIES' IN THE ART OF WAR

ANTONY CUMMINS AND YOSHIE MINAMI

Illustration on page 3: screen depicting the Battle of Sekigahara, or the Battle for the Sundered Realm, October 21, 1600.

First published 2013

The History Press
The Mill, Brimscombe Port
Stroud, Gloucestershire, GL5 2QG
www.thehistorypress.co.uk

Reprinted 2016, 2017

British Library Cataloguing in Publication Data.
A catalogue record for this book is available from the British Library.

ISBN 978 0 7509 5664 2

Typesetting and origination by The History Press
Printed in India by Replika Press Pvt. Ltd.

CONTENTS

ACKNOWLEDGEMENTS

This book could only be dedicated to Chikamatsu Shigenori himself, who, in his foresight collected the secret traditions of Iga and Koka for the education of all who follow this path.

A special thank you must be given to Kevin Aspinall who posed for some of the images and to Mr Takashi Shimizu for his aid with some of the more unintelligible sections of the text. Also, my thanks to Jayson Kane for his vast efforts with the images and illustrations and lastly to Jackie Sheffield and Shaun Barrington for their editorial help.

AUTHOR'S NOTE

> This document was written on the fifteenth day of the first lunar month in the period of Kyoho Four (1719). Also, all of the important points from Hara Yuken Yoshifusa's [ninja] scroll and the associated oral traditions have been passed on and have all been given, together with a certificate of qualification [to Chikamatsu, the transcriber of these words].
>
> *Koka Shinobi no Den Miraki*

This single passage above initiated my search for the elusive ninja scroll of the shinobi named within it. To my bitter disappointment, the shinobi Hara Yuken Yoshifusa was nowhere to be found, nor was his text, making what was a great lead for a full Koka ninja manual a dead end, or so I thought. That was until I considered that maybe, just maybe, the scroll was not under his name and that the information found within it was also recorded in one of Chikamatsu Shigenori's own scrolls. Chikamatsu Shigenori was a tactician-scholar (more on him later) who continued devotedly to record ancient and military ways, so my thought process was, *why would he not transfer this information into his own scrolls*? This led me to hunt down his more obvious military texts, the ones listed under his schools, such as *Zen-ryu* and *Ichizen-ryu*, but after scroll upon scroll and search upon search, there was nothing to be found. It was, again, with a flash of inspiration that I knew I been looking in the wrong place, I had been searching for certain titles only, so I changed my angle and found a listing of his major and extended works, all collected under his name, which contained around 150 scrolls, and lo and behold, there it was, glimmering among the black ink of the other titles – 'The Use of Spies'.

To find some of the ninja scrolls of yore, you have to put yourself in the position of a writer of the medieval period and take note that the word shinobi is not considered a positive one in the world of Edo-period Japan. However, *Yokan* – 'The Use of Spies' is a reference to Sun Tzu's thirteenth chapter in *The Art of War* and that chapter of this famous text contains the Five Chinese Spies – where there are the Five Types of Spy, surely the shinobi will follow. The reasoning paid off; document number in hand, I and my translation partner, Yoshie Minami, ordered copies of four of Chikamatsu's scrolls that appeared to have this spy theme.

Waiting in the dark streets of Warabi, Japan, I saw Yoshie approaching and we moved on to a traditional English tea shop, a local haunt where many of our translations are completed. After a pot of tea and the usual cake I could contain myself no longer and Yoshie produced the now familiar red and white envelope that a certain university send their documents in. Beaming, she said 'it is *full* of shinobi information'. Hours and more tea later, she had read through the titles and given a brief explanation of most of the points found in the manuals. I sat back, knowing the day would come – in the not too distant future – where I would be writing this introduction, knowing that my happiness would spill on to this page.

But the 'fairy tale' does not end there; the scroll quoted at the beginning of this introduction was a scroll from the area of Koka, as the words were being said by the Koka ninja Kimura. However, to my pure delight, Chikamatsu, in a wondrous turn of events, had not only recorded most of the Koka traditions, he had also studied with a master shinobi of the Iga line and recorded and compared the *two* traditions. Words cannot express the satisfaction of such a find, not only do we have a reputable figure in Japanese history, and not only did he learn and record the written and oral traditions of the shinobi, but in his foresight collected both the traditions of Iga and Koka, the premier spies of all of Japan. Eerily, in his own introduction he states that his recording of these very secret skills was for posterity and for the future study of those to come who follow the path of the shinobi, but not in his wildest dreams would he have considered that they would be here in English, the language of the world, to be studied across the globe.

It is with great pleasure that Yoshie and I present to you the collected shinobi skills of Iga and Koka and a commentary on Sun Tzu's 'Use of Spies', compiled and taught by Chikamatsu Shigenori – a ninja treasure beyond price.

Antony Cummins, 2013

CHIKAMATSU SHIGENORI

Famous for tea and the tea ceremony, the image of Chikamatsu in the west is considered from the wrong angle. When searching for information on his life, often the primary focus is put on the only other work of his published in English, *Stories from a Tea Room Window* – a text on curiosities pertaining to tea – which is often followed up with 'He was also known to be a military tactician'. This is a skewed vision of a man who can be considered one of Japan's premier Edo period military scholars who also happened to like tea. However, what is buried even deeper in the historical mist is Chikamatsu's connection to the shinobi, Japan's spies, now known as ninja.

When reading any Edo period military scholar's work on the science of warfare from Japan, you will almost always find references to shinobi, but again, almost all of the time this is a brief instruction on how to use shinobi or how to defend against them, containing no more than a few paragraphs. Chikamatsu on the other hand is a licensed master of a full Koka ninja tradition and also a student (if not a master) of the shinobi arts of Iga. Master Kimura of Koka awarded Chikamatsu a Kyojou licence, confirming that he had a full transmission of the scrolls, complete with all of the oral traditions, making him an instructor of *shinobi no jutsu* – the arts of the shinobi. However, we must not fall into the trap of placing him solely in the realms of the shinobi, Chikamatsu also had licences or instruction in many other military and classical arts, such as poetry and Shinto, and overall was a man dedicated to the study of warcraft.

According to the Shirinsokai, the official record of the samurai of Owari province, Chikamatsu's grandfather, Chikamatsu Shigehide, started his service for the third inheritor of the Owari Tokugawa clan, Lord Mitsutomo, at an unknown date before 1664. He became the magistrate overseeing two districts and held the office of magistrate of the water supply (*mizu Bugyo*), a position of importance. The Owari Tokugawa clan were one of the three great houses of Tokugawa, alongside Kii (Kishu) and Mito Tokugawa. His son, Chikamatsu Shigekiyo, who was Shigenori's father, entered into the service of the Owari clan in 1691 and later became the 'head of craftsmanship' and controlled the lumber for the lord's residences in Edo.[1] He died in 1732.

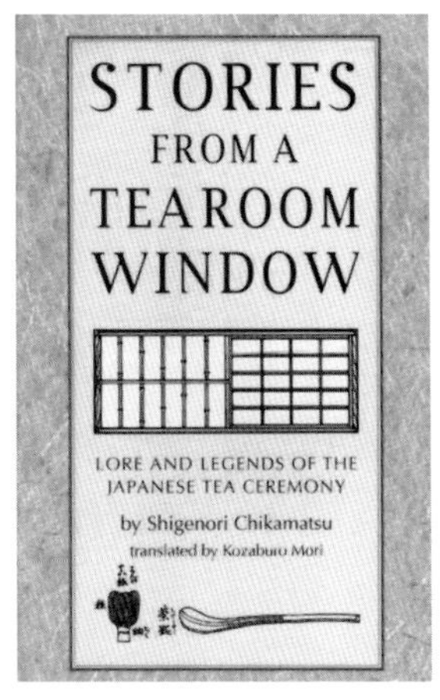

Chikamatsu Shigenori was born in 1697, the name he was most commonly known by was Hikonoshin. He was also known as Renpeido (after his training hall) and Nogenshi. At the age of sixteen, in 1712, he entered the service of the Lord Tokugawa Yoshimichi[2] as an Otoriban (a form of page or helper) and was transferred to the lord's residence in Edo in 1713. At this point he put on a demonstration for the lord and performed Iai (sword quick draw) from Katayama-ryu, also Takanao-ryu military arts, Shin'nen Ryu staff fighting, and other military ways. The lord was greatly impressed and promoted him, bringing him to his side as a close retainer. The lord had an enthusiasm for the martial arts and had the aspiration to found a school called Zen-ryu 全流 Zen meaning 'all' or 'complete'; it would be a school that integrated the best from various different schools. With this intent, he taught Chikamatsu military warfare himself and also told him to teach the military arts to the lord's own son, Sir Gorota, who would be his successor. However, the lord died that year and Sir Gorota also died very young soon after. This meant that Lord Tsugutomo succeeded as the head of the family and from then the situation went against Chikamatsu and he returned to Owari province.

In Owari he had time to spare and devoted himself to training in military affairs and founded a school called Zen-ryu (also Zen-ryu Renpei Den) which he later changed to Ichizen-ryu – presumably to continue his original lord's wishes. The school started to take students in 1715.

Chikamatsu's own writing states that he learned Naganuma Ryu Heigaku (military studies) with master Saigyoku-ken Saeda Masanoshin and was licensed (Menkyo Kaiden) in 1731. Naganuma Ryu had strict criteria for those who are licensed with Menkyo Kaiden and out of more than 1,000 students, only ten people or so were given such a qualification; Chikamatsu was one of them. It appears that Chikamatsu did not hesitate to learn from any teacher, even after he had his own students and was considered the master of a school. He compiled and wrote a vast amount of work, which he started at the age of seventeen. More than a hundred of his works are still in existence today, most of which are concerned with military studies but also the collections include considerations of the tea ceremony, of poetry and other subjects. Chikamatsu died on the seventeenth day of the second month in 1780, over the age of eighty.

Chikamatsu appears in a collection of stories collected by a group known as the Tenpokaiki 天保会記. This was a group of samurai serving the Owari Tokugawa clan, with the aim of conversing and recording things of interest, such as literature, poems, history and other various subjects. This band of intellectuals was organised by Fukada Masatsugu (1773–1850). The subjects brought forward by the members were recorded in six books between 1830 and 1843. The first meeting of the group was held in the era of Tenpo one; hence the group was named Tenpokai. Chikamatsu appears in Book Three of the Tenpokaiki records, which states:

The grave of Chikamatsu Shigenori.

> A retainer of our domain, Renpeido Chikamatsu Hikonoshin Shigenori, each morning, washed his face and hands, dressed himself in Hakama [traditional trousers] and prayed in front of the kamidana altar, which was dedicated to Amaterasu Daijingu Atsuta Hachimangu. His prayer was thus 'Please afford me success in war.' Then he also prayed to a charm from Akiba Jingu, saying 'please protect us from fires'. After this, he went down the Keikoba training room and performed five kata from each of his martial arts [bugei] and then had breakfast. He kept this routine all through his life, each and every morning without fail.

Also in the text it states:

> He was friends with Komiyama Soho[3] who he knew from Tea Ceremony. Once he said to Soho, 'I would like to ask you for a large favour. Will you please say yes?'
>
> Soho said, 'You should first tell me what it is. Otherwise, I cannot agree to it.'
>
> Shigenori replied, 'Please lend me 100 gold coins.'
>
> Soho said, '100 gold coins is not a small amount and I have to talk to my head clerk about such a matter. Therefore I cannot lend it to you at this moment.'
>
> Shigenori said, 'I do not need it now but I would like to have you promise the amount to be given upon the eruption of war.'
>
> Soho said, 'In that case, the answer is simple. I will lend you the money you require.'
>
> Shigenori said, 'I am so very happy that you have understood my situation and have agreed. I will be able to sleep soundly from tonight on.' This was said with a smile.
>
> The next day, he showed his gratitude with a present of sake and food and continued sending him sake and food at various times; such as Bon and the end of every year for the rest of his life.

I, Masanori say:

Consider the character of Shigenori with these two episodes above, and you will see that he was solely obsessed with military matters. However, he also wrote with concern about many subjects, including; the tales of Mukashibanashi (Old Stories) and other books. He liked the Tea Ceremony and wrote Sayukojidan (Old Stories about the Path of Tea), which cleared up a lot of uncertainties about that way. They say Shigenori lived in the Horeki and Meiwa eras and served under the reign of Taiko.[4]

Tenpo 10 (1839), 15th day of the first month
Fukada Masatsugu

THE RENPEIDO – CHIKAMATSU'S TRAINING HALL

The Renpeido 練兵堂 was the name of the school that housed Chikamatsu's teachings and to date we do not have a reference to its exact location, nor its size. However, we have concluded that it is Chikamatsu's personal house and that it was situated to the south of Nagoya (Owari) castle. The term Renpei is a word taken from Naganuma Ryu to mean 'training soldiers' while Do 堂, means 'construction' or 'building', but in this case 'place of'. It is unknown just how many students Chikamatsu had under his banner of Ichizen Ryu, and how many attended training at the Renpeido building. However, this is where Chikamatsu would have passed on his military knowledge, including this shinobi information.

KIMURA OKUNOSUKE FUJIWARA YASUTAKA SENSEI OF KOKA

Kimura Okunosuke was a Koka-mono, or 'man of Koka', who was employed by the Owari-Tokugawa clan. Unfortunately, his name is not in the listings, meaning that he was probably too low ranking to be named and was probably a statistic under one of the headings in the clan listings. What we do know is that he had a brother, who seems to have been more accomplished and was recorded in certain documents. The only mention we get of Kimura Okunosuke is from a listing in the late Edo period, much later than 'our' Kimura, which states that he got his name from his forefathers and that he was employed to look after cannon and to undertake construction and lastly to be a guard. While not conclusive, it would seem that this is 'our' Kimura's descendant.

Chikamatsu penned the document; Mukashibanashi 昔咄, *Old Stories about the Owari Tokugawa Clan* in 1738 and here, briefly, Kimura and his brother appear and while the reference is slight it shows that the Kimura brothers were heavily involved with shinobi-jutsu and were of course Koka-mono for the Owari branch. The first episode does not contain the Kimura brothers, however, it does deal with Koka-mono of the same domain and therefore has been included here. The episodes are as follows:

The first episode

Ueno Kozaemon[5] and his family have their origins in Koka of Omi. From old times and until his grandfather, his family were in charge of shinobi as well [as other things] for the whole of Kai province.[6] The grandfather's brother, Wada Magohachi was a minor captain of 'the fifty bodyguards' [assigned to the lord of Owari Tokugawa]. Once, Lord Zuiryu-in[7] departed from Owari to Edo and went as far as Kawasaki.[8] The lord heard that a retainer was told to keep the gates [of his castle] strictly locked when they left and he had hoped to send a letter back to the castle about a matter he had forgotten to attend to. He called for Wada Magohachi and when he came to see him the lord gave him a letter and told him to deliver it discreetly and bring back the reply to the lodgings they would be in the [next] night. Magohachi immediately went back [to the castle] to deliver the letter and came back with the reply to the lord's lodgings the very next evening. The lord was extremely pleased and praised him.

Magohachi's son, Magodayu, changed his surname to Ueno, and his son is the Magohachiro of this day [1738]. Wada and Ueno are both local names from Koka.

The second episode

Kimura Okunosuke's brother, Kimura Kogoemon, was called by the lord to a place called Yokosuka[9] and was asked to demonstrate the use of 'floating aids' (Ukigutsu) to the lord. The demonstration went very well. Also, Kimura Kogoemon was good at shooting muskets because he was studying at Tatsuke-ryu.[10] On top of this, he discussed the subject of shinobi and of his family traditions, and was thus appointed to enter the Gojunin gumi (fifty bodyguards). These were men chosen from second sons and below of good families; that is, families who at least held the position of magistrates and were known for their skill at arms. These bodyguards would keep the lord safe while on journeys or when hunting, etc. Kimura's brother later became a sub captain or kogashira, which is a leader of a squad of five of those men, making ten kogashiga in total. When the Mt. Koya incident in Kishu took place in 1692 and ronin were about to rise up united in arms, the shogunate sent examiners to decide if an army should be sent and if other ronin were going to join the rebellion. At this point, [the lord] told Kimura Kogoemon to discreetly investigate the situation. Kimura Kogoemon presented a detailed report including drawings of Mt. Koya and so on within two days [most likely through the Koka network].

TAKENOSHITA HEIGAKU MINAMOTO YORIHIDE SENSEI OF IGA

This warrior is the man whom Chikamatsu studied Iga ways under. It appears he is a man of Ise, a province near Iga, and that he is not a native born of Iga himself; however, he teaches the Iga arts and therefore we have added the suffix 'of Iga' to his name. In truth, this means that we consider him to be 'of Iga' in the sense that he is an instructor of Iga traditions and of that lineage.

In the Mukashibanashi document, *Old Stories about the Owari Tokugawa Clan*, Chikamatsu referred to Takenoshita Heigaku in an article on archery. Chikamatsu states, 'I learned the arts of how to shoot arrows from horseback – [traditions found in] Ogasawara Ryu – from Takenoshita Heigaku, who is an Inshi[11] (hidden warrior) in Yokkaichi of Ise province. The skills from Heigaku [of an Iga line decent] are transmitted in two other books. I added them to the subject of warhorses in the basic course of Ichizen-ryu, which I founded and started teaching [this archery] in 1727. We are the first to train and shoot arrows from horseback in Owari province.'

NAGANUMA RYU

Naganuma Ryu appears frequently in this book and is a school of military studies founded by Naganuma Muneyoshi (1635–1690). He studied Koshu Ryu and other various Japanese schools, also he studied Chinese manuals including the Bubishi (not the Karate text but the Wubeishi, a colossal war manual) and also the Kikoshinsho (Ji Xiao Xin Shu 紀效新書). He produced the document, the Heiyoroku, which is the main text of Naganuma Ryu, which attached importance to gunnery and the drilling of soldiers and taught more than one thousand students (Chikamatsu was not among them as he was too young). His schools were inherited prominently by two students, Saeda Nobushige and Miyagawa Ninsai, and these two branches continued up to the end of the Edo period.

SAEDA MASANOSHIN OF NAGANUMA RYU

Born in 1654, this warrior is quoted extensively by Chikamatsu in this book and was of course Chikamatsu's instructor in this school. Also known as Saigyoku-ken he was from Owari province and at first served Shinjo province and then moved to Edo where he studied and taught Naganuma Ryu. Later in 1706 he served Tsu province (possibly remaining in Edo), and died in 1742.

THE MANUALS USED FOR THIS TRANSLATION

This book is based on five manuals found in the Chikamatsu Shigenori collection, two sets of two, a main manual and its companion text, normally with extra information and further oral traditions, plus the warning scroll given by Kimura of Koka. We have decided to divide the teachings into Part I and Part II, firstly, the shinobi skills of Iga and Koka and secondly, the Iga and Koka commentary on Sun Tzu's thirteenth chapter from *The Art of War*. Even though Chikamatsu's manuals can be found in various locations, our manuals are from Hosa Library in Nagoya, details of which are given overleaf:

Original Title	Title in Latinised Form	English Translation	Text Location	Catalogue Number
用間加条伝目口義	Yokan Kajo Denmoku Kugi	Oral Traditions from Iga and Koka and Additional Articles on the Use of Spies.	Hosa Library	12–146
用間伝解口伝抄	Yokan Denkai Kudensho	Oral Traditions on the Study on the Use of Spies	Hosa Library	12–118
用間伝解	Yokan Denkai	A Study on the Use of Spies	Hosa Library	12–185
用間俚諺	Yokan Rigen	Traditional Sayings on the Use of Spies	Hosa Library	12–178
甲賀忍之傳未来記	Koka Shinobi no Den Miraiki	For the Prosperity and Future of Koka Shinobi	Mukyukai Library	N/A

THE YOKAN KAJO DENMOKU KUGI – ORAL TRADITIONS FROM IGA AND KOKA AND ADDITIONAL ARTICLES ON THE USE OF SPIES

The bulk of this translation is of this manual and is made up of the collected skills and oral traditions of Iga and Koka from the respective lines and schools that Chikamatsu studied under. Each article is given a name and an explanation of the skill is given below it, sometimes in a three-stage format. Chikamatsu himself states that he is recording these skills for posterity and that he has written down around 60–70 per cent of the oral traditions (Kuden) that were passed on along with the original manuals for generations. This leaves 30–40 per cent of the oral traditions missing and they cannot be guessed at in full. However, a clear example of these missing skills can be seen in Kimura of Koka's statement – found in the manual *Koka Shinobi No Den Miraki* – where he says that the deepest secrets of Koka shinobi include subjects such as 'Toothpick Hiding' a skill which does not appear in this manual and therefore should be considered one of the missing oral traditions.[12] That being said, Chikamatsu also states that he has tried to hide nothing and that he has attempted to be as straightforward in his description and recording as possible, so that no future misunderstanding of these skills can take place, and that while some sections are missing, it is not something to fret over.

It must be noted that while Chikamatsu claims these are the secrets of Koka and Iga lineages, we know now that not all lineages from Iga and Koka shared identical skills. The Bansenshukai manual, a collection of Iga and Koka skills, does share elements of the skills found here to quite an extent, however there are significant variations and information that is not found in the Bansenshukai is found in the Chikamatsu scrolls and vice versa. Furthermore, scrolls such as the Shinobi Hiden and the Gunpo Jiyoshu, which can also claim Iga lineages, differ somewhat from the skills listed here, showing that while many manuals claim to have the collected skills of Iga and of Koka, a unified, definable Iga or Koka line of ninjutsu is theoretical, even in the Edo period; meaning that there was no single repository of a full curriculum of the skills of Iga and Koka and that elements differ from family line to family line. That being said, it is clear that a firm and identifiable record of the shinobi arts is a reality and while skills do differ between lineages, the aims and themes of the teachings remain the same and countless times techniques appear in different formats that have a shared origin. This means that even though terms like Iga-ryu and Koka-ryu are bandied around at least as early as the 1700s, they were not a codified school with a firm curriculum and they had no central base. Chikamatsu's own school, like that of other famous schools, maintained a dojo or training hall, Iga-ryu and Koka-ryu cannot be seen in this way.

Taking all the above into consideration and understanding that the information in this scroll was compiled from the teachings of real shinobi of Iga and Koka lines, we have to let the scrolls speak for themselves, showing just how impressive and detailed the shinobi skills of these regions were, taught as Iga and Koka shinobi once taught them.

THE YOKAN DENKAI KUDENSHO – TRADITIONAL SAYINGS ON THE USE OF SPIES

This is a transcription of the above manual, the Yokan Kajo Denmoku Kugi and was based on teachings given in a lecture by Chikamatsu Shigenori who was named in the work as Master Nankai 南海先生. The information was written down from these lectures by a man named Suzuki Shigeharu. Nothing is known of Suzuki but the text has an ownership stamp naming a man called Suzuki Shinkichi, who is presumably a descendant of his. On the whole this text is identical to the first scroll as it contains the same information and format; however, at points additional sheets of paper have been over-laid on the manuscript with further teachings that contain more detail for some of the subjects. Unlike the two manuals that follow this one – which, while similar, have been kept separate in the English form – we felt that it was necessary to combine both of the above manuals to form one translation, a complete record of the skills and methods written down by Chikamatsu. Those who wish to study these two manuals as different entities will find that the position of the additional information from the second manual has been identified in the footnotes.

THE YOKAN DENKAI AND THE YOKAN RIGEN MANUALS

These two manuals have been combined into one single translation but retain their own identities. The reason for their combination is that the second manual is a copy of the first but contains more oral traditions and further explanations on each point. Therefore, in this book we have merged the two manuals together as one but identified which manual says what. The two manuals are commentaries made on Sun Tzu's thirteenth chapter in his *Art of War*; this chapter is known as *Yokan* in Japanese and translates as 'The Use of Spies'. This chapter is accepted as the single script that helped to develop spy craft in all of Asia, including Japan. The commentaries are from both Iga and Koka traditions and allow a superb insight into not only what was considered the Chinese method of spying but also the deeper skills and meanings that the Japanese added to the original

A page from the Yokan Denkai Kudensho transcription.

text. Whether these additions were originally used in China and passed down and across to Japan is unknown, but it does permit a grand tour of espionage in feudal Japan. The commentary sheds light on tactics such as acquiring Internal Spies via Converted Spies, or the combination of use of Converted Spies alongside Doomed Spies, those who were sent to their deaths. It cannot be overstated how much of a delicious and substantial find this text is, for not only is it an explanation of the deeper meanings of the now famous thirteenth chapter but it also presents to us the tactics of the premier shinobi of all Japan. The original manuals are ordered so that a single quotation is given from the chapter and an explanation and teaching is given below. In our translation we have written the quote from Sun Tzu and then identified which manual each translation comes from. On occasion, this involves a little too much repetition but not enough to warrant the merging of the two into one text. Chikamatsu wrote them at very different periods in his life and therefore it is not correct to merge them as one here. The original text included quotations from other manuals and commentaries on the thirteenth chapter; however they have been taken out as they are not directly connected to Iga and Koka traditions and rarely talk directly of the shinobi.

THE KOKA SHINOBI NO DEN MIRAIKI SCROLL

The Koka Shinobi no Den Miraki is one of the most important non-instructional ninja documents in the collection of the Historical Ninjutsu Research Team and possibly in the world of Japanese Shinobi research. Dictated by Kimura Okunosuke Yasutaka to his student Chikamatsu Hikonoshin Shigenori, it outlines the Master's fears for the ever declining Koka shinobi and predicts the fall of the shinobi with extraordinary accuracy. The reason for its importance is the issues that Kimura raises. Firstly, Kimura says that anyone under the age of seventy at the time he is writing, (born after 1650) has no connection to anyone who was living in the Sengoku Period and thus, the practical skills have been diluted and are not realistic or second nature. Furthermore, it provides an extremely rare view of the inside of a shinobi school; it shows that ninjutsu was practised in military compounds and that within these schools there were multiple levels of knowledge[13] that were kept secret from others. This document is of vital importance, showing us how shinobi were close to their lords and how they fell from prominence, and how the shinobi of Koka were one band.

DIRECTIONS AND DAYS

On page 79 you will find that the manual suddenly takes a dramatic detour from identifiable shinobi skills as it enters into *The Shinobi Hogyo no Maki*, The *Shinobi Way of Divination Scroll*. For this part of the scroll you may find yourself as a reader perplexed to the meanings and terminology and even a little taxed. This scroll lists the associations between the movement of the heavens and the Chinese dating system. It was believed in the medieval period that when the moon or a specific constellation was in a certain position then it had the power to determine if an action at a time, an hour, a day, a date or year was positive or negative. This knowledge would have been common to many and used by military groups. The aim of this scroll is to provide the trainee shinobi with a record and understanding of these associations. To aid the modern reader it is helpful to understand that the date was identified in two ways, by the twelve Zodiac animals, the Hare, Monkey and so on, and by the Ten Heavenly Stems, these are Japanese words like *Kinoto* and *Kinoe* and together they make a series of combinations that cycle forever in blocks of 60. Lunar mansions are simply a position in the sky, 28 in all, which the moon travels around each lunar month. With just this information it is possible to understand the main aim of each entry in the scroll and what it is used for. A modern diagram is supplied to help you follow the instructions in the scroll, as often the text will say, 'count clockwise', or 'move six signs from the Monkey', or something to that effect. This you can do for yourself with the aid of the image provided in that section. The actual instructions in the scroll are ambiguous; the function is to find a positive heavenly time to perform a certain task. Therefore, while this section may strain your attention, it has been kept in the scroll for completeness and for your future study as your understanding of *shinobi no jutsu* increases. The more 'mainstream' shinobi instructions start again on page 94.

MEASUREMENTS

All measurements are given in their original Japanese form, such as Shaku, Sun and Bu. The western equivalents are easily obtained and a reduced table is given below.

Shaku	30cm
Sun	3cm
Bu	3mm
Ryo	15g
Momme	3.75g
Kin	600g
Sho	1.8 litres
Go	180ml

ARTS, SKILLS, TRADITIONS AND PRINCIPLES

Most of the skills found within this book, as in many scrolls from medieval Japan, come with an accompanying title and all of them contain suffixes that follow a generic form. Each of the differing suffixes are actually very close in meaning and are often interchangeable, however, they do still contain subtle changes. We have attempted to unify the translations of each of these to promote conformity, however, the 'feel' of them in English can sometimes not work well and alternatives have to be used. In general, we have tried to stick to the following translations where possible.

ノ傳 – no Tsutae has been translated as 'Tradition of'
之事 – no Koto has been translated as 'Art of'
大事 – no Daiji has been translated as 'Principle of'
ノ術 – no Jutsu has been translated as 'Skill of'

While consistency has been strived for, variations in this rule may occur, therefore, it is beneficial to familiarise yourself with the Japanese form of these endings.

CONCERNING IN AND YO

Throughout this book, you will come encounter the concept of In and Yo and many teachings and instructions where Chikamatsu uses one of these as a prefix or suffix. The concept is not altogether alien to a western audience; we understand it as Yin and Yang, the eastern notion of Dark and Light. While the ideograms remain the same in both Chinese and Japanese, the pronunciation changes to In 陰 and Yo 陽 in the latter.

This idea of dark and light is integral to the Japanese system of thought and has a strong presence in the text, however a direct translation changes depending on the context and the grammatical structure used. This translation has kept the original reading of In and Yo as we feel that a western reader will benefit more from this use of the original language instead of a series of differing translations.

ENTER THE REALM OF THE SHINOBI

One fundamental mistake most people make when reading shinobi texts is that they class the translations as a full curriculum, a complete set of skills. In one respect this is true, that is the translations are full shinobi skill sets, but those people writing these manuals and carrying out these tasks had other skills as well. These shinobi were a part of the warrior culture and you should understand that the skills of the ninja were an add on, used to enhance a warrior's already formidable skill-set. In the Sengoku period – the time where these arts were perfected – a samurai would be a proficient fighter, horseman and sometimes, a war master, but more than all these things, such as swordsmanship, archery, swimming, sea warfare, survival techniques, they had *shinobi no jutsu*, for those special needs – making them warriors with that little extra, exceptional advantage, an advantage that could almost guarantee employment and was kept very close and secret.

Therefore, before you finally reach the text, take a moment to stop and consider where you are about to journey. Deep in the mountains and war-torn lands of the Sengoku period, war-hardened men were operating, where many killed and were killed. They did this with the knowledge you are about to read ingrained in their minds, memorised through years of training; and it was a daunting task to put these skills into operation in a land riven by conflict. Therefore, while these translations may simply be of great interest to a modern audience, they were once the deepest of secrets and repositories of applicable knowledge.

NOTES

1. These positions appear to be menial to western understanding, however, they are positions of authority in the lord's house.
2. The fourth inheritor of the Owari Tokugawa clan (1689-1713), his reign was from 1699-1713, he was also known as Enkaku-in – his Buddhist name.
3. A famous merchant in Owari, who was very successful in the medicine business.
4. Tokugawa Munekatsu who reigned from 1739 to 1761.
5. Ueno Kozaemon was a Metsuke, in this case the supervisor of retainers and later promoted to a close aide to the sixth lord of the Owari Tokugawa family, who reigned from 1713–1731.
6. Most likely when Tokugawa took over Shingen's forces.
7. Tokugawa Mitsutomo, the second lord and head of the Owari Tokugawa family.
8. The distance from Nagoya (Owari) to present-day Kawasaki city is more than 430km and therefore it appears that the episode is referring to a place called Kawasaki in Owari province, which is located no more than 70km from Nagoya. The story is about the speed with which he performed the task, which he most likely did on horseback.
9. Yokosuka is a place in Owari province (present-day Aichi) where Lord Mitsutomo had a residence.
10. One of the three major schools of gunnery for the Tokugawa shogunate.
11. This term possibly means someone who used to be of the samurai class and now is of upper peasant position, above the other three classes but below full warrior status.
12. This skill does appear in other manuals.
13. This should not be taken too literally, the author is showing depths of understanding which are kept secret, not a grading or curriculum.

PART I

THE SKILLS OF IGA AND KOKA

用間加条伝目口義

YOKAN KAJO DENMOKU KUGI

THE ORAL TRADITIONS OF IGA AND KOKA AND ADDITIONAL ARTICLES ON THE USE OF SPIES

The Upper Scroll

THE TRADITIONS OF IGA AND KOKA

There are two scrolls[1] of additional articles pertaining to Koka shinobi and have been transmitted [to me] by Yasutaka [Kimura Okunosuke sensei]. Also, there is one scroll pertaining to the traditions of Iga shinobi which has been transmitted [to me] by Yorihide [Takenoshita Heigaku] sensei. For all of these manuals there is a small amount of oral tradition [that accompanies them] and these traditions have been passed on by memorizing and learning them each by heart so as to not lose this knowledge and therefore, since ancient times, they have never been written down.

I [Shigenori] have now made a resolution to combine the following two documents together; the Kajo [Koka traditions] and the Denmoku [Iga traditions] and also I have collected and written down six or seven out of ten of the oral traditions which have been passed on. The remaining three or four oral traditions out of every ten are such deep secrets that they have been omitted here, this is because they should be taught only through hearing them thoroughly and directly and only from one person to another. Therefore, you should have no doubts about these omissions but instead devote yourself to the study [of the arts which have been recorded here]. I have no intention of making anything secret within these writings, but only wish to pass down [these traditions] to those who have the ability and ambition [to learn them] so that these skills will not die out. This is my true hope and anyone who aims to learn these ways should learn them with extreme diligence and without neglect.

Written In the first days of the third month of 1737
In Bishu of Owari province
By [the founder] of the Renpeido school of war
Fujiwara Shigenori

1

忍之訓傳

Shinobi no Kunden

THE TRADITIONS OF THE MEANING OF THE IDEOGRAM SHINOBI 忍

Koka traditions say:
This ideogram is called shinobi as they act in secret,[2] away from the ears or the eyes.

Iga traditions say:
When you make a thing [that you have in hand] a deep secret, it is called shinobi.

The deepest secret traditions of Koka say 忍 is from the ideogram 堪忍 which means 'patience'. Thus it is called shinobi because the [skills of the shinobi] are skills you need to endure the unendurable or any difficulties or hardships that are imaginable. Those who follow this path should think of what the name of their profession truly means. If you keep this in mind at all times, you will be able to complete shinobi missions and distinguish yourself.

I, [Shigenori] say:
The point taken from the deepest secret traditions above is entirely appropriate.

2

忍之起原

Shinobi no Kigen

CONCERNING THE ORIGIN OF SHINOBI

Iga traditions say:
According to the upper scroll of Records of Ancient Matters,[3] the god Susanoo changed Princess Kushinada into a magic comb and put the comb in his hair, this comb was called Yutsutsumagushi, next [Susanoo] exterminated Yamata no Orochi, a giant eight-headed snake. This is the very origin of shinobi no Jutsu, the arts of the shinobi.

Koka traditions say:
According to the lower scroll of the same writing as above, the god Takamimusubi sent a bird called Nanashi no Kigishi – which means 'a common pheasant of no name' – to investigate [how the god he sent years beforehand was behaving on earth]. This is the origin of the shinobi.

Yasutaka [Kimura of Koka] says:
Our shinobi originated in the Age of the Gods, and their skills have been passed down until today, even across this extensive period of time. Although it is rare [that lineages should last so long], this actually is the case [with ours], this is because our shinobi [arts] are military skills which have their roots in the divine and are not [only] human affairs,[4] as are those military tactics in other countries.[5]

I, [Shigenori] say:
I think the tradition most appropriate is the one which says its origin is the myth in which the god Susanoo transformed [the princess before he killed the giant snake]. This is because, the bird Nanashi no Kigishi, that is, the *nameless pheasant*, was shot and killed with an arrow and therefore this theory is not appropriate. Furthermore, it is also inappropriate to say that someone from low birth[6] who has no name was the ancestor [of the shinobi].

One night, Master Yorihide [of Iga], said to me that the Emperor Jimmu and the god Yamato Takeru both fought crusades and achieved great military feats by following the divine strategies of the god Susanoo. Susanoo began using shinobi no jutsu and killed the serpent in that story with ease. Therefore, his divine strategy gave birth to military tactics which have endured from then until now and here, in this point, is found the essence of divine warfare. However, these days shinobi no jutsu is thought to be performed by those who are middle level samurai and even lower people,[7] that is, people feel it is a job for those [samurai] of lower position and also they read the ideograms 竊盗[8] as 'shinobi' and call [shinobi] yato-gumi 夜盗組 which means 'a band of night thieves'. These points are totally incorrect. Some military scholars teach that shinobi no jutsu is not good enough for practical use, but this is because they themselves have not achieved a mastery of military tactics, therefore, do not be swayed by their shallow opinions.

The god Susanoo about to join the giant serpent in combat.

3

風二乗ルノ傳

Kaze ni Noru no Tsutae

THE TRADITION OF RIDING ON THE WIND

Koka traditions say:
The wind has no shape and neither do the shinobi. Whatever the situation is, you should change your appearance dependent on the circumstances at hand and then infiltrate.

The traditions also say that the ideogram for wind also makes the word 'fuzoku' 風俗 which can mean 'manners and customs'. Take note that manners and customs vary in 66 ways in 66 provinces.[9] Even in the same province things can differ; remember, even those areas on the east side and the west side of Lake Biwa have different manners and customs from each other. Basically, it is primary to know what the manners and customs of any given area are like and to imitate them as much as you can, and to not act against the ways of local people.

Iga traditions say:
As the wind blows in through even a smallest gap, shinobi can infiltrate through any such small breaches. However, if the place is perfectly guarded and has no openings, it may be the case that not even the wind can blow in at all. Therefore, if there is no gap, you should not try to infiltrate needlessly, for if you do and if you push yourself too much, it can cause mistakes in the end.

Secret traditions say:[10]
Wind never shows its shape to people's eyes and likewise, shinobi no jutsu should change so it too will not be detected.

4

雲行之傳

Unko no Tsutae

THE TRADITION OF THE MOVEMENT OF CLOUDS

Iga traditions say:
To communicate an urgent message from where you are on a covert mission, you should arrange for Tsugi Hikyaku, *relay messengers*. These messengers should be arranged at intervals and they should relay the message by running from one person to another.

In later times[11] it has become Haya Hikyaku *express messengers*, but there is a way to do this discreetly and this is only a way to be used in peacetime. You should have commoners instead of Bushi warriors[12] perform this and pretend that the purpose [for their movement between areas] is to communicate current market prices; this is done so that people will not identify their real purpose [of passing messages and gathering information].

Koka traditions say:
Where you can see each other, use flags to communicate and where you cannot see each other, use smoke signals. Arrange for the number of flags, or the number of smoke signals to be made in order to communicate your message. At night communicate to each other with lights. For all situations, you should arrange messengers at intervals just as is done with relay messengers, so a message can be relayed as far as 100 *ri* and in no time at all.

Secret traditions say:
An example – if there is a roadside shrine on a hilltop, you should arrange for what specific message will be intended when you will put a flag on the right side or what meaning is carried if you put the flag on the left side. If you use lights, arrange for what message you will communicate if a fire is in front of the shrine and what message if the fire is in the rear of the shrine. This is also the case when you use a large tree instead of a roadside shrine. In any case, you should experiment and test before you actually use it.

5

蚤虱之傳

Nomi Shirami no Tsutae

THE TRADITION OF FLEAS AND LICE

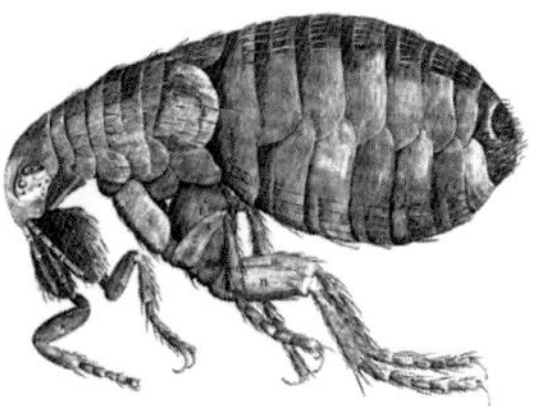

Iga traditions say:
Fleas and lice live on people and go with them wherever they go, no matter if it is a palace or a magnificent castle. Just as fleas and lice move around, shinobi no mono should first make connections with certain people and then infiltrate by accompanying them. Take note, you should only infiltrate through a route where people do not pass, that is, by crossing a moat, or crossing over a wall [by infiltrating in stealth only] when it is a state of emergency.

The first thing you should do is pass through a gate by sticking close to people. If it is impossible to pass through a gate, then you should infiltrate by crossing over a moat or by climbing over a wall.

When[13] teaching fundamental lessons, if you draw an illustration of a castle and show it to your students and ask them to identify the best point of entry for infiltration of that castle, you will find that those who are not talented tend to obsess with the name [and common image of the] shinobi and will say such things as: 'the corners of the wall are good for infiltration as they have good footholds' or 'I would climb up the concertina-style wall.'[14] A smart student will say that he would go through the gate where people are passing through anyway. You may think it is difficult to go through a gate which is known for being secure but it actually can be easier [than infiltrating by stealth]. If someone says he would go through the gate, then you can say he is talented and has an aptitude for this job. Such a method of infiltration sounds difficult but in fact, it is actually easy, because you pass by disguising yourself as someone who can actually pass through there with ease.

Also tradition says:
If you use tens of shinobi no mono, there is no place you cannot infiltrate. A small number of fleas or lice can be got rid of, but once they have increased in number, it is difficult to kill them all as they are hiding everywhere. In the same manner, if tens of people infiltrate in various ways through various places, they will succeed in the end in collecting information by listening and watching.

The deepest secret traditions say:
Fleas or lice stick to the body inside of the clothes. In this manner, make anyone, if they are serving the enemy clan, or if they are a peasant or commoner, follow and obey you, make a shinobi agent for your side [from the enemy] and send them out – this is the so-called Internal Spy. If you make an ally of one of the enemy in any shape or form, and have the person bring you in [to their area], then you can infiltrate anywhere, no matter how deeply you need to go, and you can get secret information from anyone, no matter how close it is kept. It is very difficult for the enemy to defend against you if you gain an ally within them.

6

見詰聞詰之大事

Mitsume Kikitsume no Daiji

THE PRINCIPLE OF MITSUME AND KIKITSUME

Koka traditions say:
Kill two dogs while they are mating, take the vagina and the male genitalia and put the second one into the first one and bury them both at a crossroads. Next pray every morning for seven days while offering a bowl of water, asking that your wish will be fulfilled. Also spend these seven days without speaking a word and make a vow of silence. After these

seven days take both of them [out of the ground] and put them in your house. Pray for a further seven days just as above while offering a bowl of tea each time. After this, dry them both in a shade, then put them on a Goma altar and burn holy sticks of invocation for seven days. When finished you carry them in your clothes and when you perform Mitsume (observing) and Kikitsume (listening), you hold them tightly in your hand.

Mitsume means observing thoroughly to discover something and Kikitsume means listening intensely with the aim of discovering information.

Also traditions say:
These skills are not seeing or listening casually, but watching or listening closely and carefully.

Secret traditions say:
On top of seeing or listening closely to discover something, you should go further by observing or listening in more detail, to find out the very reason behind the surface of affairs and to discover truly why a situation is like it is. You should listen or see intensely, speculating on the external facts to penetrate to the truth deep inside.

Some traditions of the shinobi[15] have three-fold teachings, that is, they have three different levels of the art [from lower to the higher principles]. As mentioned in the document Miraiki, it is extremely important that you should not teach everything at once to a given person.

The lower tradition states:
You should start with 'light' teachings and observe the character and nature of the student thoroughly before you admit him into the higher traditions.

The middle tradition states:[16]
You should not only see or listen to things in a superficial way but observe and listen thoroughly with special care, this is known as Mitsume and Kiktsume.

The upper and secret tradition states:
About things you see or hear – you should not take this issue lightly but try to observe or listen deeper to get the truth of things, considering the reason for why any situation is thus, this is Mitsume and Kikitsume. You should not fail to catch the truth when it surfaces, even if it is only visible for a moment, you achieve this by observing and listening, and speculating on what people are thinking beneath the surface and consider their future actions.

7

蜘蛛之傳

Kumo no Tsutae

THE TRADITIONS OF THE SPIDER

Koka traditions say:
Spiders cannot function without their web, just as a shinobi no mono cannot cross over a wall or roof without rope. This set of traditions has various skills and [only] two of them are described here.

To cross over a high wall, observe the wall from the outside and throw a Ban'no[17] tool (grappling iron) over what you consider[18] to be a bracket of the wall. If the [Ban'no tool] has been hooked [over the bracket], tighten it by pulling down on it and climb up using the rope. When you climb up onto the roof of the wall, descend to the inside with the same rope again. This is just like a spider as it crosses over.

If you have infiltrated in an unfamiliar place or you are not sure of how to return, just leave a rope attached to a post or tree where you have just entered and go on ahead while holding onto the rope. If you do this and if it is difficult for you to return, you can find your way back by tracing the rope back to your original location. This is the same as spiders as they always go ahead with thread wherever they go.

Secret traditions say:
Whenever you infiltrate, you cannot enter without a starting point. *Traditions of the Spider* is about seeking a start point for your coming infiltration. When spiders build their web, they need a start and secure their thread to build their web. When you infiltrate a castle or mansion, according to the traditions, you should make a connection with someone inside and have the person let you in.

8

大忍之大事

Ooshinobi no Daiji

THE PRINCIPLE OF THE GREATER-SHINOBI

Iga traditions say:
In order to secure a successful outcome for your shinobi, you should put your men into various provinces and make acquaintances at times of peace, do this so that you can communicate anything with each other freely. Such a plan will not work well if you try to place your contacts [in enemy territory] when emergencies have already arisen. Therefore, to do this, you should [infiltrate your agents] by utilizing the *way* of the arts, be it; poetry, Renga verse linking, tea, the *Go* game[19] or other artistic hobbies, making acquaintances in various provinces and having [the agents'] names known widely and by many. Also you should use anything that is popular at the time and will let you have contact with various provinces easily.

[There is an additional note here that is too faint to read]

According to Master Saigyoku [of Naganuma Ryu], you should make acquaintances with someone serving each of the various clans.

Oral traditions state:
For example, if you have solid acquaintances [from Owari] up to Kuwana of Ise but you have none in Yokkaichi, then you should have an acquaintance in Kuwana help you make acquaintances in Yokkaichi. By doing things like this, you can develop connections in various places.

9

陰陽忍之大事

In Yo Shinobi no Daiji

THE PRINCIPLE OF IN AND YO SHINOBI

Koka traditions say:
There are three layers of teachings for this subject to be transmitted directly and they will be passed down separately [outside of this scroll]. There is also a deep secret for this and it is written as follows:

In wartime, boundaries are strictly guarded and they check people so strictly that it is difficult to send your spies into the enemy province. On the other hand in peacetime, it is easy to have those from your province move and live in other provinces, therefore, you should have your spies go to other provinces and settle there. However, if you have an emergency, that is a sudden turn of events before you have set up your spies in the enemy area, you should examine those peasants, craftsmen or merchants as well as bushi from your own province who have children or brothers living in enemy lands. If one is among them, then make the father or brothers [from your province] communicate to their children or brothers [in the enemy province]. As you have these people in your lands you can use them as hostages and use [the relatives in the enemy province] as you wish.

10

山彦之傳

Yamabiko no Tsutae

THE TRADITIONS OF ECHOES

Iga traditions say:
Echoes here means; hearing news that confusion has taken place in another province. Upon such news, you should send more spies to get further information.

Also the traditions say:
To get your shinobi to infiltrate from the west, you should make noise in the south and attract the enemy's attention to the south, which makes it easier for you to get in from the west. But take note, if you make noise in the west and infiltrate in the east, then this is not a good strategy.[20] Therefore, you should make noise in the west and infiltrate from the north.

Koka traditions say:
It is called the art of Echoes because you gain someone in the enemy and correspond with that person.

11

火フリ火トホシノ大事

Hiburi Hitoboshi no Daiji

THE PRINCIPLE OF FIRE HANDLERS

Iga traditions say:
Fire skills are essential for night raids. You should have those who handle torches go in first. They are called Hiburi (or Hitoboshi). They do not have their lights seen by the enemy until they have infiltrated secretly and only at the last moment do they provide their allies with fire (torches or grenades) and keep throwing them to burn down enemy quarters. The enemy have to deal with two things at the same time, which are defending against your attempts to burn their position down and the advance you are making

through fighting fiercely and cutting through them. Therefore, you should choose an appropriate person to lead the [fire] troop. In older times they used to be called Hiburi no Oyakata – Captain of the Fire Handlers.

12

水月之大事

Suigetsu no Daiji

THE PRINCIPLE OF THE MOON ON THE WATER

Iga traditions say:
If the enemy strictly guards his boundaries, it will be difficult for you to send your shinobi there but still they normally always send their [agents] to your province. If an enemy spy 間 infiltrates your side, you should not defend against him but instead let him in and use him as a Converted Spy. This is just like when the moon is in the sky, it reflects on water, so this is called the Principle of Suigetsu 水月 *The Moon on the Water*.

Deep secrets say:
The moon reflected on water can be seen but actually has no form,[21] therefore in reality you cannot take a hold of it. The enemy spy is exactly the same as this, and it is extremely

difficult to obtain him so securely and, therefore, be able to use him as your own spy. You should be fully aware of this and know it is essential to win over his heart with careful planning. The art [needed to achieve this] varies depending on the situation and the person, so details are not written down here.

It[22] is essential to discover thoroughly what the character of the enemy spy is like and to make him feel intimate with you. You should treat him with kindness, give him money, pay for his clothes and food, making him feel obliged to you and talk to him about what is beneficial for him and what is not. By doing this you will be able to convince him that he is doing the right thing [by changing sides] and make him divulge his real intentions and the truth so as to use him for your own purposes. You should not tell him what your plan is beforehand as you should construct the plan dependent on the needs of the time and the situation and arrange it considering his mind-set. If you fail to gain his heart for certain, you will make mistakes when the time of implementation comes. Therefore, be sure not to fall into the difficult [idea] of grasping the reflection of *The Moon on the Water*. This fits in with Master Saigyoku's notes.

13

四知之傳

Shichi no Tsutae

TRADITIONS OF FOUR WAYS OF KNOWING

Koka traditions say:
To make a Converted Spy, you need to thoroughly know the mind of the enemy spy who has been sent to spy on you. Otherwise, it is difficult to turn him to your side. There are four ways of knowing his mind, and these four are exactly like the method of the doctor.[23]

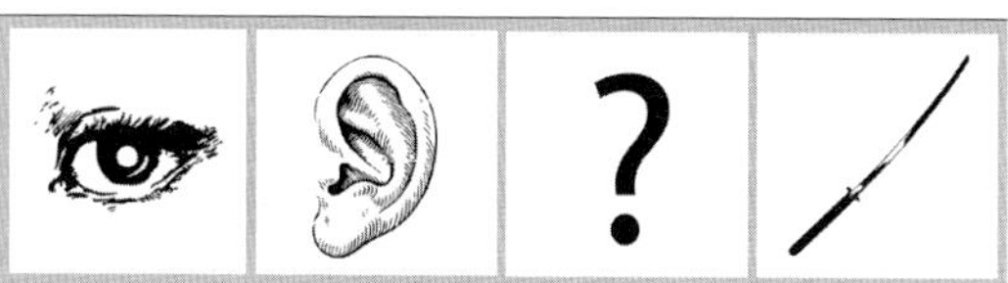

1. Observe 望
2. Listen 聞
3. Question 問
4. Cut 切

The first way:
Observing – this is to observe the person's manners and customs and way of behaviour.

The second way:
Listening – this is to listen to him talk about various things to know their mind by what they say.

The third way:
By questioning – this is to ask him about what he likes to know his words and deeds.

The fourth way:
By cutting – this is to give him a test to observe the way he talks and behaves so that you can discover his true mind, which may have been hidden from you.

If you follow all of these above measures, you will – without fail – find out the real intentions and feelings of the enemy spy. Once you know everything about him thoroughly, you will be able to figure out a way to use him as a Converted Spy.

14

占合之大事

Ura-awase no Daiji

THE PRINCIPLE OF SECRET CONNECTION

Koka traditions say:[24]
Sun Tzu says you should furnish your spy and the enemy spy with false information so that they will provide the enemy general with incorrect material and force [the enemy side] to make wrong decisions. However, Sun Tzu does not describe the art in detail, therefore you should make a Doomed Spy by considering this Principle of Secret Connection as described below.

Koka traditions say:
There has been no way to prevent spies from meeting and communicating with each other in either Japan or China at any period of time. Therefore, you should be very selective about knowing if your spy is good or bad. If your spy is better than the opposition's spy, he can get detailed information about the enemy and speculate further on that information, while pretending to give the opponent information about your side but in fact actually giving away nothing. This is called Suri to Suri no Deai 摺ト摺ノ出合.[25] Sun Tzu transmitted the art of Doomed Spies, but this art has not been fully studied in Japan, thus it is difficult to carry out such a plan by using the skill, because of this I am going to explain this art in detail here. Master Saigyoku's [of Naganuma Ryu] teachings are exactly the same as this *Principle of Secret Connection* which is from the Koka traditions, so you should learn both ways. First you should make one of your own spies angry and have them form a grudge against you and then let him know of a [falsified] secret plan. Your grudging spy then will tell an enemy spy of this plan and the enemy spy will repeat this to his own lord. However,[26] if your spy is faithful, he will not voluntarily give anything away to the enemy side, but as he speaks his mind he will leak the information in the end.

Looking at this in detail, Sun Tzu and Master Saigyoku's teachings say you should first make a false plan and then let your spy know it and have him form a grudge against you. However, Koka traditions say that you should force your spy to have grudge against you and *then* let him know of your secret plans. You should be aware that the order has been reversed.

The enemy lord will hesitate to take action even with this new information. Therefore [to ensure he does not hesitate] you should let someone who has been sentenced to death, a prisoner from a previous war or a criminal who is in prison for committing a serious crime, overhear the guards or officials who are on duty talk about the [fake] plan. If they overhear it, they will think they have obtained quality information and want to escape any way they can and inform the enemy lord of what they know, so that they can claim a reward. At this point you should pretend to let him escape as if it was by mistake.[27] The runaway will return to the enemy side and inform the lord about the same plan [as the spy]. As the information matches perfectly with what the spy has said before him, the enemy will believe it and decide to take actions to deal with the matter. You should then outwit them by taking actions to deal with the move the enemy will make and defeat them.[28] As a result of this, they will be badly defeated and the enemy lord will become angry and kill the spy and the runaway prisoner who brought the misinformation. This is what Sun Tzu called Doomed Spies.

15

三合ノ大事

Miawase no Daiji

THE PRINCIPLE OF THE THREE-WAY CONNECTION

Koka traditions say that this art is nearly the same as the above *Principle of Secret Connection*.

Iga traditions say:
This is almost the same as the above principle but with a minor difference. The difference is that you should first make a plan and broadcast an image of trying to keep it secret, but in truth you allow it to be given away by 'mistake'; to a page, woman, monk or anyone else who is chatty, so that it spreads as a rumour and your spies and the enemy's spies will know of it. In this way it will end up that the enemy lord on hearing it will *defend* against the information even if he does not believe it completely and does not take a *positive* action. Therefore, when the plan is actually conducted, the enemy will have the advantage over you and your side will be at a disadvantage, this is because they are prepared beforehand. However, in order to win a substantial victory at a later time, you have to endure this disadvantage and leave the enemy pleased with their temporary gain. Then, perform the above mentioned *Principle of Secret Connection* and you will gain a massive victory in the end. In this skill, you put together everything from the three stages mentioned above and that is why it is called *Three-way Connection*.

After the enemy have obtained benefits from this ruse two or three times, they will rely on nothing other than what that particular spy says. Then, at this point if you let that spy know something that may deliver them a massive victory, they will be eager to take hold of such an opportunity. When performing the art of the Doomed Spy like this, you should give them these small victories at first and if they have fallen into the plot, win a massive victory over them [with false information]. In this art you should first construct a [fake] plan and leak it out so that the enemy will hear of it. Next you should act in accordance with the plan and allow the enemy to gain benefits, then use Ura-awase no Jutsu – *The Art of Secret Connection* – and let a prisoner or criminal escape 'by mistake', so that he will go to the enemy with the information of your [false] plan. The enemy take action based on this false information and in return you can engage your true plan and strategy. In this way you will win a total victory. This is called *The Art of Three-way Connection*.

Mori Motonari[29] fought with Sue Harukata several times, but whenever they fought Harukata had known Motonari's tactics and Motonari lost every time. Because of this Motonari took notice of a blind entertainer who recently took up service with him and Motonari wondered if he was informing the enemy of their tactics in secret. He tested his theory a few times and decided that the blind man was actually a spy for Harukata. Motonari was pleased to discover this fact and one day he set up a secret meeting with his close retainers [including the blind monk], and said, 'If Harukata attacks us in Itsukushima, we will be ruined,' and with this he sighed. The blind man heard this and told the enemy leader, Harukata, the full story. As Harukata had benefited from the information given by the blind man several times, he was very pleased to hear this most important information and attacked at Itsukushima immediately. Motonari was truly joyful at this because the enemy had fallen for his plot and thus he decided to strike at Sue's forces with a night raid. Just before the enemy forces landed on the island of Itsukushima, Motonari had the blind man taken in front of him and he said: 'We are going to take Harukata's head today thanks to your efforts.' Then he had the blind man drowned in the sea and went on to win a massive victory. This is exactly what *The Art of Three-way Connection* is about, though the people involved were not aware of this fact.

16

一離之傳

Ichiri no Tsutae

THE TRADITION OF SEPARATION

Iga traditions say:
This art is to know that if an enemy is to use a spy, then you should expect it and deal with the spy properly. You should consider [the future] by placing yourself in the enemy's position.[30]

If your army has won a victory, thanks to the information brought by a spy, even if a small victory, you should not rely solely upon the words of the same spy in the next battle. It is even more so if your spy tells you something that will apparently bring you an important victory and also be careful if this information matches with that which is told to you by someone who has surrendered or a refugee, or someone from your side that was captured by the enemy and has escaped and returned. Therefore you should not take advantage of this information and detach yourself from [the apparent benefits] and stay separated from it. For example, even if the information you have obtained seems to be certain and will enable you to capture the enemy general and burn the enemy castle down to the ground, you should not use it and you should let it go. If you do not use the information, the plan the enemy have constructed will not only fail but it will turn out that they will lose heavily because they tried to use the skill of utilizing a small defeat to win the war.[31] On top of that, they will begin to have doubts among each other and consider the possibility that someone within them may be playing both sides and secretly communicating with the enemy. That will make them hesitant and suspicious in making plans and cause them to lose in battles to come.

17

正合之秘術

Seigo no Hijutsu

THE SECRET SKILL OF CORRECT CONNECTION

Iga traditions say:
If you use the above tradition of 'Separation' and detach yourself in the manner described above and just let [the enemy win so that they feel they have the upper hand], then know that it can sometimes be harmful to your side, so much so that you cannot allow it to happen. However, if you use the supplied information, it may turn out that you become trapped in the enemy plot.[32] In such a situation, you should use *The Secret Skill of Correct Connection*. This means you should stay firm, with enough preparations to guard yourself whether the information is true or false, making sure you will suffer as little loss or harm as possible. You should concentrate your efforts on damage control instead of trying to find victory. Remember, you can be trapped within the enemy plot because you hunger for victory or to gain profit. If you detach yourself from [your desire to] win, there is no way that you can be trapped.

For example, if you are informed that you will suffer a night raid from the front and the rear, you should prepare defences not only in the front and rear but also all the four directions and stay firmly guarded. Then if the enemy do in fact come to attack, you should fight to defend against them. If the enemy run away, having lost, you should not follow them but stay back and maintain tight defences. You want to follow them simply because you want to win. The enemy intention might be one of the following three:

1. While displaying a desire to attack from the front and rear, they will in fact break through from the sides.
2. They will lose and run away intentionally. If you pursue them then they will have ambush soldiers set up to defeat your troops following them.
3. Also, if you do pursue them, running after their troops, they will capture your camp or castle with separate forces [which have been held in reserve for this very purpose].

However, none of the above schemes will work against you successfully [if you use this art].

Also, if you receive information that someone [from your own force] is betraying you, you should capture the person and put them in prison, but meanwhile try to bring the whole plot to light by talking to people and probing into the facts to gain an understanding. Then, in this way, if it is false information, you will discover the truth before long. If you fall into their trap and kill the falsely accused, he cannot be resurrected. Or when you are informed that the enemy are going to use the Revolving Formation,[33] you should not divide your army to deal with it accordingly, you will have fewer numbers at hand, so if the enemy attack where you have fewer numbers you will be defeated [this is the main aim of their strategy, to divide you]. When you hear that they are going to use the Revolving Formation, pretend to act and comply with the information In Yo – [by that it means only a surface or act for show] but actually by using the principle of In – [that is to be deceptive] – arrange it so that the number of men that you *pretended* to divide [in a false display of separation] will actually converge and attack the enemy from the rear when the correct signal is given by you. This is called the *Art of Seigo* or *Corresponding Appropriately*. This art should be varied according to the situation.

As[34] is shown above, in *The Art of Separation*, when it is found that what a spy and a runaway prisoner have said correspond with each other and it becomes obvious that the enemy has sent a Doomed Spy over to your side, you should not use the information but separate yourself completely from it. However, dependent on the situation, it can be disadvantageous if you totally avoid the matter and you feel that you cannot just leave the issue because the information may be true, though it seems that if you use it you may fall into the enemy plot. In such a case, you should use *The Art of Correct Connection*, that is, you should deal with the situation by taking appropriate measures, preparing yourself with patience without seeking for [immediate] benefit and just deal with the situation with a calm and collected mind and in a composed manner. You will suffer an unexpected defeat if you try to obtain excessive benefits and if you are too anxious to gain victory. If you have acquired information to attempt to foresee the enemy's next move, you should prepare your forces as much as possible just in case the information is true, but not in a way to put yourself in a disadvantageous position, this all depends on if the information turns out to be true or false. In this way, you should make arrangements so that you will gain a victory if the information is true and also, you will be in an advantageous position if the information is false. Furthermore, you should not overstretch yourself in an attempt to gain victory but detach yourself from the possible benefits gained and instead keep your mind serene and retain a firm guard, so that an enemy plot will not succeed. For example, if you are informed by your spy that you will suffer a night raid from the front and the rear, you should prepare defences not only in the front and rear but all four directions and stay firmly guarded. You should not think too much of winning but secure your defences, focusing on supporting those areas [under attack that need relief] and stay well ordered in your position. If the enemy does come to attack you, and as you have taken measures to deal with it beforehand, they will not be able to defeat your forces. When the enemy withdraw, you should not follow them without care, restraining yourself from seeking victory but instead staying firmly guarded. [In this case] the enemy tactics may be: pretending to attack you from both sides and retreating, which will be followed by an attack from the left and right sides. Alternatively, they just wish to attack your forces outside [of the fortifications] and try to lure you out while they retreat and another ambush troop will attack and capture your camp, or they may ambush and attack the soldiers who have come out of the fortress. If you are careful not to be too anxious or to try to gain a quick victory and instead stay firmly guarded, then all these above methods will not work. Also if they have a plot and have a spy or prisoner tell you that someone from your side may betray you or turn traitor and that they are playing both sides, then you should not kill the supposed traitor immediately without careful examination but instead just secure him so that he cannot flee. While he is secured you should investigate and try to find out the true facts of the matter, the truth will surface before long. Whether it is real treachery or an enemy plot, you should not kill him without care but stay calm and act thoughtfully, so that even an enemy stratagem will end up becoming advantageous for you. Be sure not to be trapped easily but take your time, making sure to do things step by step and thoughtfully.

If the enemy have a plot and try to tempt you out of your position and then attack you by ambush or if they try to take you by surprise, then you should pretend to have been trapped by their ploys. One way to do this is to keep your forces as a whole, but secretly prepare separate units elsewhere. If you find that they actually have laid ambushes, you should command [some or all of] the hidden units immediately to fight together according to the situation. Alternatively, you should divide your forces openly and pretend to be trapped by the enemy plan but should arrange for signals. When they think that you do not have enough forces – because you have divided your army – and they become confident that you have been trapped, when they advance, attacking you, getting even as deep as your close command group,[35] then you should engage the rear troops that you have hidden and attack the enemy from the front and the rear with all your forces put together. This will bring you a massive victory, and this art which is taught in the tradition of Iga fits in with our traditions of Naganuma Ryu.[36]

Form your army openly (Yo) but divide it discreetly (In), and also divide it openly (Yo) while bringing it together discreetly (In). Note: you can divide your forces [secretly] while not dividing your [observable] formations. If done this way it is highly reasonable to say that you will win both in reality and in that which is inferred.[37]

Secret traditions to enlighten your awareness and wisdom:
After you have broadly studied *shinobi no jutsu*, you should then learn from those families of Iga and Koka as they have family traditions in Iga and Koka that stretch back for generations. You should learn the true traditions of the arts.

18

内之氣外之氣之傳

Uchinoki Sotonoki no Tsutae

THE TRADITIONS ON INTROVERT CHI AND OUTGOING CHI

Koka traditions say:
There are two kinds of chi; *Introvert* and *Outgoing*. An example of this is: when you have infiltrated a target house as planned and without incident, you will be strong minded, like iron and so determined that you will be in a state of mind which enables you to smash anyone who fights against you, tearing them into pieces. This is when your chi is strong; which is reflected in the common saying:[38]

'Let them all come – you can bear any arrow, shot or shield'

On the other hand, when you have completed your job and come out in the garden to retreat, that strong chi will be weakened and all you can think of is to flee as soon as possible before someone has the chance to detect and follow you. If you are followed in such a state of mind, you will be easily killed without being able to retaliate. You should think about this and be fully aware that your chi may be weakened or distracted in the manner as is mentioned above. This is exactly how your chi works.

19

内之氣外之氣訓傳

Uchinoki Sotonoki Kunden

THE TRADITIONS ON INTROVERT CHI AND OUTGOING CHI

Iga traditions say:
These traditions are about when you have completed your mission of infiltration and are in the process of retreat. [The name for Outgoing chi is] 'Sotonoki' and has similar pronunciation to 'Sottnoku' [which means 'retreating quietly']. The sounds 'Ki' and 'Ku' are interchangeable because they belong to the 'Ka' column, that is, Ka Ki Ku Ke Ko [in the Japanese system].

When in the state of Introvert Chi, you will normally hold yourself back from fighting and try to escape as quietly as possible whilst also trying to make no noise. However, in your mind you should be determined to fight your way out if you are chased and attacked. You should continue with this state of mind, so that you will not go through the change between Introvert Chi and Outgoing Chi.[39]

For example, if you have dogs barking at you, and if you drive them away and are relieved that you are rid of them, you will find that soon they will follow you and bark at you again. On the contrary, if you think you should have killed them and you wish they would come back and bark at you again so you can in fact kill them, it will actually turn out that they will not follow or bark at you. You should be fully aware of this.[40]

20

小城之傳

Kojiro no Tsutae

THE TRADITIONS OF 'SMALL CASTLES'[41]

Koka traditions say:
When you are on the defensive, you should hang up a paper mosquito net[42] and sleep within it. If there is a small hole [in the side] you can see outwards from inside of this net but note that you cannot be seen from the outside unless someone gets close to the net and peeps in. Therefore, when you come out to fight, you can tear the net off to prevent it from getting in your way. It is quite different from a cotton mosquito net in this way.

Iga traditions say:
If you find paper mosquito net(s) set up when you have infiltrated [a position], you will have difficulty in approaching the target as it allows them to see from inside to the outside with ease. Also, it prevents you from having a good aim at the target with a shorter bow[43] or short musket, so that [you] will be held back [because of uncertainty]. Therefore, in older times, someone who had an enemy who sought to take their life in revenge used to hang paper screens up [as they slept], even in winter [when there are no mosquitos].

21

要之大事

Kaname no Daiji

THE PRINCIPLE OF THE PIVOT

Iga traditions say:
A pivot is what holds together ten sticks to make a fan and allows it to serve as a fan. Tens or even hundreds of spies will all solely depend on the commander-in-chief. You should understand this by remembering the following; when those generals who were serving the Founder of the Han Dynasty in ancient China were not achieving their aims, it was only Chen Ping 陳平 who was executing his tasks properly.

Also shinobi-headmen[44] are most important. Although [it is thought that] shinobi talk directly to the lord in most cases, it is only the headman and those who are most skilled [who have direct contact with him]. If there are tens or hundreds of people, they cannot work very well without a good headman, so an appropriate person should be chosen for the task [as commander of shinobi]. You should be aware of this, thinking a fan does not serve without the pivot – as if there is no pivot then all ten sticks will fall apart.

Surviving spies are most important – take note that you should not employ someone unless he has such a personality and qualities as is explained in the study written by Master Saigyoku. Know that, if there are ten people and they are different in ten ways, you should keep a number of spies and employ each by what he is good at to fit in with your needs at any given time. The person in command of all those people should know how to use spies thoroughly and use them according to their abilities; this is called *The Principle of the Pivot.*[45]

A pivot is what puts together ten sticks to make a fan and allows the fan to serve as a fan. Likewise tens or even hundreds of spies will all solely depend on the commander in charge of the men. The commander cannot fulfil their aim unless he knows everything about everyone of his men and use them appropriately. Remember,[46] although there were many generals serving the Founder of the Han Dynasty in ancient China, only Chen Ping knew the arts concerning the use of spies. As mentioned previously, it was often the case that shinobi had the chance to talk with the lord directly, but not all of them had such a close relationship. Therefore, it is important that a commander of shinobi knows the arts of the shinobi extremely well and his men in a proper manner. There is another issue to take into account, that of using and being used, which is related to this art. For information, refer to the final chapter of scroll Ninjin Miraiki – *The Mind of Shinobi and a Future Account.*

22

面討之傳

Omote Uchi no Tsutae

THE TRADITIONS OF ATTACKING THE FRONT

Iga traditions say:

To infiltrate secretly and kill the commander-in-chief, the rule is to divide your men into three attack groups. The first unit is Hiburi (also Hitoboshi) which are Fire Handlers, the second unit are Omote Uchi and they attack the front, and the third unit are called Wakekiri and attack the rear. Attacking the front means to fiercely attack with swords from the front and so you should choose intrepid and hard warriors for this job. While they are fighting fiercely the Fire Handlers throw fire and the Wakekiri Rear Attack Troop come from behind.

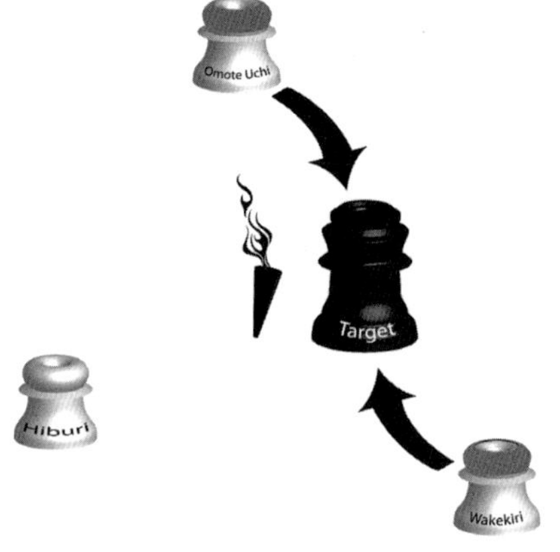

23

ワケキリノ傳

Wakekiri no Tsutae

THE TRADITION OF THE REAR ATTACK TROOP

Iga traditions say:
Wakekiri is to go through the rear gate and attack the enemy from behind. For this task, you should divide your men into groups of two or three people. When you attack in groups you should attack each single enemy with these three men, and if there are two or three enemy men, six of your men should attack them.[47] By using this method you can secure a victory.

24

松明之占手之大事

Taimatsu no Urate no Daiji

THE PRINCIPLE OF DIVINATION BY TORCHES

Iga traditions say:
When you need to capture or kill someone and it is of great importance and if the target is a good fighter, you should throw large torches around his position. You should throw three torches from three directions to smother the enemy with the fire and smoke so that you can capture them while they are stifled and confused. There are traditions which came later on where poison was embedded in the torches, however, as the fight will take

place at close quarters, it will do harm to your men too, thus you should not use this way any more.

The word Urate [from the title] does not mean Urate [a rear direction] but means the way of divination and the three torches should be named as follows:

1. The first torch is for the God of War 軍神祭リノ松明
2. The second torch is for the God of Luck 時ノ運祭ノ松明
3. The third torch is for [your] general's life 大将壽命ノ松明

You should choose the best warriors and have them throw torches and attack with the intent to capture the target. In all probability you will succeed with only the first or the second torch and before using up all three torches. If you capture the target with only the first torch, you should cry out aloud:

軍神ノ守リツヨシ
Gunshin no Mamori Tsuyoshi
'We have full protection from the God of War'

If you succeed on the second torch you should cry out:

時ノ運モヨシ
Toki no Un mo Yoshi
'Fate is on our side'

These things are meant to infuse your men with spirit.

If things are not going very well and you think it is too risky for you to make a raid on the target on that night and you decide not to attack, then it will discourage your men. If this is the case, you should first try this way of throwing torches and tell your men that you should call off the attack and retreat, saying that the divination turned out to be unlucky for your side.[48]

If this art of Urate is performed after the three steps of the Fire Handlers, the Front Attack Unit and the Wakekiri the Rear Attack Unit, by the time you are using the Principle of Divination by Torch, you should be completely sure of your success. This art is performed in order to rouse your men enormously and make them spirited.

This series of Fire Handlers, Front Attack Troops, Rear Attack Troops and Divination by Torches is found in a Noh theatre play called 'Kumasaka Eboshi Ori'. In ancient times, even killing and robbing used to be done in large groups. They usually had a commander and everyone worked under the commander's order and mostly all was done by night raids. Night raiding is the best way to attack a large number with a small number, which was noted by [the famous warrior] Tametomo, who was a part of the Shirakawa regents house.

Shinobi no jutsu has specialised in night raiding since ancient times and thus, these skills have been transmitted. This is why, in other military schools, the ideograms 竊盜 are used to mean 'shinobi' and it is hard to stop them from using those ideograms in this way.

It is even more important that you should not hand down these skills without full teaching and not give them away to others.

In later times [which came after this ancient period], there was a skill to be used when capturing an *important* criminal, this was where you lit long torches[49] and thrust them at the criminal from every side so that they blind the criminal and so that you can immobilize them with the three tools [of capture, which are the Kumade rake, the Tobiguchi rake and the Sasumata two-pronged pole arm]. This technique was also successfully developed by being based on the art of Fire Handlers and their torches.[50]

25[51]

内ノ風外ノ風ト云事

Uchi no Kaze Soto no Kaze to Iu Koto

THE ART OF INTERNAL WIND AND EXTERNAL WIND

Iga traditions say:
External Wind is a term that stands for those who attack while *Internal Wind* represents those who are the defenders. Being strong 強ミ or weak 弱ミ is known as the following; Hayashi means to be quick and Ososhi which means to be slow. These terms are idiolects that are used in the schools or families of shinobi[52] and are not known to others.

These words mentioned above also come from the play [*Kumasaka*][53] *Eboshi Ori* [as talked about earlier]. Furthermore, [in shinobi jargon] the ideogram for wind 風 means 'manners' or it can mean the status of the [enemy] at that moment. As for being quick or slow, this implies a connection with that passage from Sun Tzu about being as 'quick as the wind'.[54]

26

心結之大事

Kokoro Yui no Daiji

THE PRINCIPLE OF CONNECTING WITH THE HEARTS [OF OTHER SHINOBI]

Iga traditions say:
In order to prevent your shinobi no mono from allying themselves with the enemy and becoming a Converted Spy, you need to form a very close relationship with them and lead them to have gratitude towards you and to fill them with appreciation towards you until they love you to the core of their bones.[55] Make it so it is as intimate as the relationship between father and son, or brothers. What you should keep in mind is the need to create

a relationship between two unrelated people and make it as [strong] as it is between father and son. Remember, you should try to understand the minds and the thoughts of your spies in all cases.

It is very dangerous if you are unsure that you have secured a relationship with the person in question. As the name of the art implies, unless you are sure that you have captured them absolutely in every respect, then you should not use them. You should always make it a principle to make a close relationship with the person in question as if you are father and son, even though you have different surnames. It is essential to have them comply with you by conducting yourself based on the principles of heaven.[56]

無刀傳

Muto Den

THE TRADITION OF NO SWORD[57]

[Supplement to the above.]

He, [the shinobi] is normally presented to the lord in person, and has the opportunity to talk in close proximity, as the saying goes, 'directly from the mouth to the ear'. It is expected that at such a time [the shinobi] will not be allowed to wear a sword. However, he may hide a stabbing blade[58] 刺刀 in his clothes or he may snatch the sword that the lord is wearing on his waist, with the intent to kill him. Therefore, you should never let your guard down and always remain vigilant.

Even if it is the case that [the agent] has been acting as a completely faithful and single-minded retainer, you should not fail to take proper care and assume that he may have an ill intent towards [your] lord and try to kill him.

27

重播之大事

Shikimaki no Daiji

THE PRINCIPLE OF DOUBLE SOWING

Koka traditions say:
You should continue to give your spies twice or even three times the amount of money that the enemy would bribe them with. If this is done, your spies will not take the risk of allying themselves with the enemy for the sum of, say, 100 pieces of gold[59] but instead they will choose to serve you for the amount of 200 or 300 pieces of gold as it is easier [and more profitable] to do so. They will only pretend to be Converted Spies [for the enemy] and obtain evidence that they have been offered 100 gold coins and show this proof to you. This situation will benefit your plots and help you achieve great victories. If the enemy fail every time they try to acquire Converted Spies, then they will give up trying to make Converted Spies in the end. On top of this, the enemy spies will envy your own spies as your spies are more highly rewarded and therefore, enemy spies will easily yield to and obey you. Next you should offer [the enemy spy] 200 koku[60] in salary, which the enemy spies can keep, as well as the 100 koku paid by their own side and if this is done they will surely serve as Converted Spies for your forces.

This art is meant to achieve the following two things: to prevent your spy from becoming an enemy Converted Spy and secondly to make a Converted Spy of an enemy spy. This art is called Shikimaki, double sowing, as you sow money double [the amount paid by the enemy].

28

草苞之大事

Kusazuto no Daiji

THE PRINCIPLE OF THE GRASS BUNDLE

Koka traditions say:
It is said that a domain will decline because of 'grass bundles', that is, a domain will decline because of money and bribes. [The name originated] because in wars of the past, there were lots of Tozoku thieves and people carried money in grass bundles so as not to be robbed.

Connected to the saying that 'a domain will decline because of grass bundles (money),' you must remember to have spies act to the fullest [regardless of the money it takes]. You should not spare expenses but give liberally as much money as your spies wish, even to the levels of thousands of gold coins, and let them spend the money freely dependent on their judgement and you should not ask them later how the money was spent. Even if spies are given a thousand pieces of gold, they spend them in such various ways, ways which will not allow them to enter every transaction into ledgers. If they try to do so, they will not be able to work very well as a spy. At times you may spend 100 gold coins and yet not yield any worthwhile results, while at other times you may obtain ten times the value expected for 100 gold coins, even though you only spend a quarter piece of gold.[61] Also, you may need to involve yourself in activities of all kinds, such as going to plays, taking a female or male prostitute for pleasure, gambling, composing poems or taking part in a Renga verse linking party, also kouta ballad singing, Joruri dramatic recitation and koto or shamisen musical instrument gatherings, among other things, most of which are activities that you can in no way enter into an accounting ledger. Therefore, since ancient times, spies have not had to perform the task of entering every expense into these ledgers. If they are forced to enter every expense, then it is a primary rule in the traditions of spies to refuse to serve as a spy. Unless both the lord using the spies and the spies serving the lord think nothing of money and spend money like water without care for how it is spent, great results cannot be achieved.

Traditions [from Koka] also say:
When you enter expenses in an accounting ledger, you have to put exactly what the money was spent on and if the expense was appropriately used or not. If you put such details in the record books then your activities will be overtly displayed, which will let people guess the operations that you have conducted. People may doubt if the money was appropriately used or not and you will sometimes be suspected of having spent money to make profit for yourself or to fulfil your own desires. Therefore, whether it is necessary or not, you should give notice [to who accounts for these things] that, if you are given 100 pieces of gold, you will spend them all, if given 1,000 or even 10,000, again you will spend them all within a short time of being given them, and [make it clear] that we [shinobi] do not care for accounting and this should be made clear before you go out for a mission as a spy. If you are not given permission on this point, then you should not serve as a spy, this is *the* primary rule for spies.

In ancient China, Chen Ping served Emperor Gaozu as an overseer of the main force and supervised every general. The Emperor brought out 40,000 kin of gold and gave it to Ping and allowed him to use the money freely without questioning how he used it. With the money Ping sent out Converted Spies to the Chu regime and he employed rare and ingenious strategies on six occasions and pacified the country.

Ikeda Sanzaemon Terumasa had his favourite retainers, Wakahara Ukyo and Nakamura Tonomori, and he had them give rice or gold and silver to help ronin[62] from various provinces. They helped those ronin who were talented and who had achieved. Hundreds of people were helped and how this rice and gold or silver was used was passed over [and was not questioned], they were not asked to record exactly how they spent the rice and the money.[63]

Oral tradition states:[64] Securing the Source by Destroying the Ends
These two people in the above case were killed in the end. Therefore, spies should be very careful concerning keeping things secret and not to give away anything to anyone. You may feel that it is too harsh to kill anyone who has heard such a secret, however, you should know that this is because if spies give away any secrets, the person who has heard it may not spread this to others but in fact they may tell someone who is higher ranking and report the spy's discrepancy. Therefore, you should kill anyone who has heard a secret, this is done to make [the spy] careful and so that he will not want to disclose anything to anyone.[65]

The Art of War States:

> It is owing to his information, again, that we can cause the Doomed Spy to carry false tidings to the enemy.

There are oral traditions about this matter and it was annotated by our Master but I [Shigenori] say you should [use the way of Doomed Spies] from the Iga traditions. This art should be the basis for a great victory but is not mentioned by Sun Tzu nor explained by Master Saigyoku. And thus this is where the Iga traditions are more detailed.

More from the oral traditions of Iga:
Be sure that you do not use this plot repeatedly and let years pass after it has been used [before you use it again]. If it works well [the first time], the enemy will be cautious so this is why you should not use it again for quite some time.

Oral Traditions from Koka [concerning the above Sun Tzu quotation]
Some skills used in battles from older times have not become common knowledge, meaning that they will still work very well [nowadays] and without difficulty, ten out of ten times, therefore you should use these any time a battle takes place. However, as an old saying goes, 'do not use a massive bow to kill a mouse', therefore, you should not use this art to win only a small battle but instead it should be used to win a massive victory.

Sun Tzu states:

> Of old, the rise of the Yin dynasty was due to Yi Zhi who had served under the Xia. Likewise, the rise of the Zhou dynasty was due to Lu Ya who had served under the Yin. Hence it is only the enlightened ruler and the wise general who will use the highest intelligence of the army for purposes of spying and thereby they achieve great results. Spies are a most important element in war, because on them depends an army's ability to move.

This is not about an art for spies itself but about what a lord who uses spies should do. Know that the lord should not use people only for their original tasks. You should understand this and remember that Yi Yin and Jiang Ziya [of ancient China] were great spies as well [as ministers].

These days great clans hire skilled Koka no mono[66] for the job of gunnery and give a little amount of money to [their] families who live in their homeland [of Koka]. The man is given a certain amount of money very discreetly every year and sometimes takes leave and time off for visiting his homeland but in actuality he travels to another place. Nobody knows which province he goes to or what he does there or anything about his task. There are lots of such people in various clans. This is a common way to use spies these days. According to Master Yorihide [of Iga], most generals of today are used to the easy and tranquil life, they are not so interested in preparing themselves for warfare and study not the writings on military science. Thus they are not acquainted with the chapter on The Use of Spies written by Sun Tzu, so they do not have enough knowledge on how to use spies. Meanwhile, spies of the present time only think of how to make a living or how to make profit for their own benefit but do not care about the standards of their family traditions. Nor do they have foresight on whether or not spies will be needed in a case where an emergency could arise on the morrow, but what they do care for is to provide for their wife and children and make an easy life for themselves. Thus, this is not the correct way that a lord should use spies and equally, how spies serving a lord should not conduct themselves. As a result of this, spies are now spies in name only and it is certainly impossible for them to function practically as agents.

Those generals who want to make good use of spies should use them as in ancient ways, while spies should follow their family traditions so that they will attain great feats. Alas, it is truly regrettable that although the deep secrets of using spies have been transmitted to this day, there is no general who can practically apply those secrets, which prevents any outstanding achievement from being made. You should be thoroughly aware of this and make every effort to distinguish yourself with astonishing loyalty and exploits.

29

蚊蠅之傳

Ka Hae no Tsutae

THE TRADITIONS OF THE MOSQUITO AND THE FLY

Koka traditions say:
Mosquitoes and flies make no distinction among those of both high and low social standing, be it aristocrats or plebeians, but approach all and everyone. [Similarly] shinobi no mono should become mixed with any kind of people; from aristocrats to beggars or outcasts with the aim of gaining intelligence, and make plans based on the intelligence they have obtained. You cannot acquire correct intelligence unless you associate with anyone without distinction of age or sex, or which of the four classes that they fall into.

Iga traditions say:
Mosquitoes gather in swarms and harass anyone, anywhere. Swarming flies, if driven away will fly away but will soon return. Shinobi no mono should come together or go away [freely] exactly in this way. Take note, this depends on you having preparatory meetings to make everyone in your team understand which measures and what systems are in place and to definitively arrange any signals.

30

知情之大事

Chi Jo no Daiji[67]

THE PRINCIPLE OF UNDERSTANDING THE CHARACTER OF PEOPLE

Koka traditions say:
The character of people varies according to which province they are from. It also depends on the sex or age of each individual. If from a large province, [personalities will] vary even within the same domain. You should be aware of this and find the differences yourself. Also, you can acquire a general idea of what the character of a province is like if you read the Jinkokuki[68] record. The information written in there is correct almost nine times out of ten. However, primary sources should be obtained by travelling around the provinces and associating with people.

31

七字之大事

Shichiji no Daiji

THE PRINCIPLE OF THE SEVEN EMOTIONS[69]

Koka traditions say:
You and your opponent both have the seven [emotions] and you should be aware of them when you construct plans:

1. Delight 喜
2. Anger 怒
3. Sorrow 哀
4. Pleasure 楽
5. Love 愛
6. Evil [intent] 悪
7. Greed 欲

Iga traditions say:
You should know [these emotions] by observing yourself[70] and always think of this principle of the seven emotions. When you are to perform your job, you should know [the limits of] your own bravery and capabilities and also consider your age before you do anything. Especially, when you have an intimate relationship with influential people, you should pay a lot of attention to this point.

THIS IS THE END OF THE UPPER SCROLL

The Lower Scroll

[The first section contains the teachings of] Master Yasutaka [Kimura of Koka][71]
[Each article is made up of] three-fold traditions.[72]
Kajo加条 – Additional articles [on the traditions of Koka].

1

木ノ葉カクレノ事

Konoha Gakure no Koto

THE ART OF LEAF HIDING

Lower tradition:
On the fifth day of the fifth month, take the liver[73] of a black dog which has no white hair. Dry it in the shade until the seventeenth day of the eighth month. Trace the Sanskrit below [on the liver] three times a day [during the drying period], then grind it and wrap it in seven layers of gold brocade and carry it with you.

Write [shown left] on the liver.

When you need to hide, hold a leaf aloft and crumple the powdered [liver] on to the leaf by twisting between your fingers and then stay still, this is done so that you will not be seen by others.

Middle tradition:
When you infiltrate a castle or mansion, you should hold a small shield toward the direction where the enemy may come from, in case arrows or bullets are shot.

Upper tradition:[74]
This art is to hide yourself in the thick growth of a forest.

2

柴隠レノ事

Shiba Gakure no Koto

THE ART OF BUSH HIDING
THIS IS ALSO CALLED GRASS HIDING[75] 草葉カクレ

Lower tradition:
Take the liver of a live otter and then take the skin of a mouse. Dry both of them in the shade and grind and mix the same amount of the two together. Blow it towards the enemy [when you wish to hide].

If you need to hide, chant[76] the Wisdom Sutra[77] three times and you will not be seen by others.

[Trace] the spell [left] once [on a leaf].

Write the spell on the above grass leaf and hide in a field, this will also help you not to be seen by others.

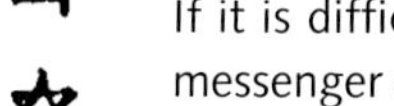

Middle tradition:
If it is difficult to get [your spy] in to an enemy province, tactically send a messenger and have a shinobi no mono accompany this messenger. Also, to be disguised as a merchant is another way to do this.

If you need to have the person remain in a province – as when the group comes back, the number of people is to be checked [at the checkpoint] – then you should take the following measures:

Have one person feign illness and pretend to die of that illness, then bury him.[78] When night comes, dig him out and refill the ground so the spy can stay in that province without being noticed. If the body is to be cremated, then you should kill someone from the [enemy] province and replace your spy with this dead body before it is cremated. This technique is also called Leaf Hiding.

Upper tradition:[79]
This art also means hiding yourself in thick grass in a field.

3

狐狼ノ傳

Koro no Tsutae

THE TRADITIONS OF THE FOX AND THE WOLF

Lower tradition:
If you find checkpoints newly built on the boundaries between provinces and it is difficult to get through them, you should take a mountain path where foxes and wolves pass.

Middle tradition:
Just as foxes or wolves deceive people by transforming themselves, you should disguise yourselves in various ways, copying the way of the people to get through [checkpoints] unquestioned.

Upper traditions:
It is difficult not to be seen by people during daytime, therefore, infiltrate at night time. Just as foxes and wolves are nocturnal in their habits as well.

4

牛馬ノ傳

Gyuba no Tsutae

THE TRADITION OF CATTLE AND HORSES

Lower tradition:
Disguise yourself as a cattle or horse dealer and infiltrate.

Middle tradition:
If the province is cattle dominant or if it is horse dominant, then you should have [your spy] disguise himself as a horse or cattle dealer and visit the province every year so those in the province will get acquainted with him and let their guard down where he is concerned.

Upper tradition:
Horse dealers or cattlemen always travel around various provinces, especially horse dealers as they get close to various samurai. Therefore, you should choose appropriate people out of them according to their ability and use them for shinobi purposes.

5

佛カクレノ事

Hotoke Gakure no Koto

THE ART OF BUDDHA HIDING[80]

Lower tradition:
You should talk to monks in various provinces to collect information about their domains.

Middle tradition:
Choose those monks from your province who are smart and have aptitude and send them out as shinobi. As there are temples of the same sect in other provinces, it is easy for them to infiltrate other places.

Upper tradition:
If it is difficult to enter another province, disguise yourself as a monk and pretend to be a pilgrim or a priest who roams about soliciting funds for a temple, infiltrate in this way.

6

穴蜘蛛地蜘蛛ノ傳

Anagumo Jigumo no Tsutae

THE TRADITIONS OF THE PURSE WEB SPIDER[81]

Lower tradition:
If it is difficult to cross over a wall, you should dig under the wall and pass through a hole.

Middle tradition:
You should not concentrate on only one side [of a position], but walk around the place several times to find out the best place from which you can infiltrate. Just as a spider builds a web, you should make arrangements [from multiple angles] before you infiltrate.

Upper tradition:
If a position is strictly guarded and difficult for you to infiltrate, you should make a tunnel, but note that you have to start digging from a distant place. This method is old and a traditional one but it is time-consuming and a drain on manpower. Also you have to have a place to dispose of the soil from the hole you are digging. Unless it is near the seashore where you can throw debris away into the sea or you acquire a huge mansion, where you will start digging this tunnel, then this method does not work.

7

霞之大事

Kasumi no Daiji

THE PRINCIPLE OF MIST

Lower tradition:
Bury two live sparrows of each sex in the ground and then dig them out after one hundred days, grind them, make a fan and put the powder between the paper[82] and carry it with you. During the hundred days, go to the place where you buried the live sparrows every day and chant the name of Amida Buddha[83] one hundred times every time you visit. If you wave this fan to beckon the enemy, then the powder will bring mist upon them.

Middle tradition:
[Ingredients]

- Blue vitriol – 2 ryo
- Pepper[84] – 2 ryo
- Japanese pepper[85] – 2 ryo
- White mustard – 2 ryo

- Soot – 2 ryo
- Chili Pepper – 2 ryo
- Ash – 2 ryo
- Salt – 2 momme 4 bu
- Powder of a clove – 1 ryo
- Sand – 5 ryo

Crush and grind the above into fine powder. Cut a cylinder of bamboo into a four sun length, and put the powder mixture into it. Paste paper over the opening and drill a number of holes before you use it. Make sure the enemy is on the leeward side of you and sprinkle it [on the wind] towards the enemy. You should be careful not to be leeward of it yourself.

Upper Tradition:
When it is misty all over, you cannot see things clearly. Therefore, when visibility is impaired and it is hard to find out what is going on, you should take advantage of any chance and infiltrate.[86]

8

犬カクレノ事

Inu Gakure no Koto

THE ART OF DOG HIDING

Lower tradition:
Pull out the eyes of a black dog, dry them in the shade for 100 days. Wrap them with red cloth and carry them. When you go on a covert mission, you should chant the following chant 100 times:

'Abira unken'

Cut the Kuji [nine-lined] grid once [in the air] and infiltrate with your thumbs kept inside your fists so that you will not be noticed by people or by dogs.

Middle tradition:

Trace the above ideogram [in the air] three times when facing a dog. Put [the upper right hand] dot [of the ideogram] in line with the dog's face[87] [this will mean that you 'write' this dot on its head] three times.

Another tradition[88] *says that you should chant the following poem three times*:

> 我ハ虎イカニ鳴トモ犬ハ犬　獅子ノ歯カミヲ恐レサラメヤ
> 'I am a tiger. No matter how hard you bark, a dog is just a dog. How could it not fear a tiger's bite, as he is the king of the beasts.'

Next, fold your fingers of your right hand – one by one – starting from the thumb and placing them into your palm while saying 'Dog, Boar, Rat, Ox, Tiger', and then dogs will not bark at you.

Upper tradition:
When you go to where there are dogs, you should take wrapped roasted rice that has [unknown[89]] mixed in with it and packed tightly. If dogs come out, you should throw it to them and have the dogs eat it. If a dog eats one, it will die.

9

飛鳥之傳

Hicho no Tsutae

THE TRADITION OF FLYING BIRDS[90]

Lower tradition:
If you jump from a height to a lower place, you should have the feeling as if you were going to jump up again. Do this at a short height above the ground.[91] This makes you land without much noise or any blunders.

Middle tradition:
Wear something such as a Tobiguchi capturing tool[92] or [wooden] staff in the same way as you would wear a sword [at your waist], but with about one shaku[93] of it coming out above your obi belt. Jump down holding the end tightly to your breast, so that it will hit the ground [first and reduce the impact].

Upper tradition:
If you do not have such a thing as described above, you should just jump down wearing your sword at your waist [and use the same skill]. To aid you, you should cover the end of your sword scabbard with metal.

10

呼出シノ術

Yobidashi no Jutsu

THE SKILL OF LURING SOMEBODY OUT [BY MAGIC]

Lower tradition:
When you need to lure somebody out, you should chant the following poem three times, and then stroke the teeth of a boxwood comb across the left sleeve of your kimono.

ソンケントチキリシ事ヲ忘レテカ其イニシヘノ藤ノ一本[94]
Sonken to Chigirishi Koto wo Wasureteka Sono Inishie no Fuji no Hitomoto

Also, you should make a powder of the charred leaves of the Kudzu vine, one which is growing upon a stone monument, then scatter the powder on the road where the opponent is expected to come from.

Middle tradition:
In the above poem, the symbols at the start ソンケン are read as 'sonken' but this is a mistake and should be read as ソムカシ 'somukaji' [which means to 'not betray']. This comes from a secret tradition. However, it has been passed down in the above way, so it appears that it has been passed down by chant in this incorrect way, therefore, it is said you should chant it in the way that people do now [and not in the original and correct way].

Upper tradition:
To lure someone out, you should make up a fake messenger pretending to be from someone in his family and who lives in their home village, asking the target to come out. Then after drawing out your target you can fulfil your aim.[95] Therefore, you first have to get thorough information about your intended target. If not, your plan will not work. A stranger can in no way lure him out because he has nothing that the target will care about. Therefore, you should work out a plan where the target will be lured out, without fail.

11

寝屋ノ大事

Neya no Daiji

IMPORTANT POINTS CONCERNING SLEEPING QUARTERS

Lower tradition:
Trace the three ideograms below on your pillow and then trace them on your futon as well.
Next, before you sleep, chant the following Buddhist incantation so that you will awake when an emergency arises:

ヲンヲノリヒシヤチイヒラホシヤラトリホトホトニホシヤラホニハンクキツリヤウハンソモカウマトロムソ若ヤマコトノ事アラハヲトロカセミヨ我枕神
Wo n wo no ri hi shi ya chi i hi ra ho shi ya ra to ri ho to ho to ni ho shi ya u ho ni ha n ku ki tsu ri ya u ha n so mo ka u[96] – *Madoromuzo Moshiya Makoto no Koto Araba Odorokasemiyo Waga Makuragami*

Middle tradition:
If you want to wake up at a certain time, you should chant the following poem three times:

ほのぼのとあかしの[浦の]朝霧に嶋かくれ行舟としそおもふ[97]
Honobonoto Akashino [Urano] Asagiri ni Shinakakure Ikufune toshizo Omou
When a day is breaking at the shore of Akashi, I am thinking of a ship moving through the morning mist which is hiding the islands.

You should chant the above poem with the word 'Ura [no]' taken out. At the same time concentrate on what time you wish to wake up. Then go to bed and by doing this you will be able to wake up at that very time you thought of.

Upper tradition:
When you need to stay on guard all night, tradition says, that you should stay up without sleeping. It is not every night that you have to be concerned [if someone will infiltrate], so [as you cannot continue to stay awake constantly] you should sleep during the daytime and remain awake at night. Furthermore, eating something little by little or having lots of strong tea will also help you to stay awake.

12

不凍大事

Kogoezaru no Daiji

HOW TO PROTECT YOURSELF FROM THE COLD

Lower tradition:
When you are on a covert mission during a bitter-freezing night, mix garlic and clove oil together, kneed it and then apply it onto your hands and feet.

Middle tradition:
If you do not have the above two ingredients, you should apply sake wine [onto your feet and hands] and rub your hands together again and again. Also, it is better to keep your thumbs inside your fists.

Upper tradition:
Collect the leaves off the stem[98] of a rice plant and crumple them into a lump. Carry it with you in your fists so it will warm you greatly.

13

角ノ犬ニモ悪マレヌ事

Kado no Inu nimo Nikumarenu Koto

THE ART OF AVOIDING BEING HATED BY 'DOGS OF THE CORNERS'

Lower tradition:
When you climb up a wall, you should climb along the internal or external corners, so that you will have better footholds. Remember, the corners are what you should pay special attention to. This art is called 'how to avoid being hated by the dogs of the corners'. This is a kind of secret language.

Middle tradition:
'Kado' from the title [in this case] does not mean 'corner' but instead means 'gate' this is because it is best to infiltrate through a gate by disguising yourself as someone who is actually allowed to go through the gate. Therefore, 'how to avoid being hated by the dogs of the gate' is a secret language or kind of idiom. Generally, most people think you should cross over a moat or climb up a wall when you infiltrate, but such things are only exceptional measures shinobi take in emergencies. Therefore, the basic rule is to get in through a gate if you can. Furthermore, if you show a drawing of a castle to a beginner and ask him how he would infiltrate the castle, and if he answers he would get in there through the gate, know that he is very resourceful and you should appreciate his talent.

Upper tradition:
This warns you that you should be careful so as to not be hated, even by dogs. It is even more important [not to be hated] by people. It is of primary concern that you should be careful not to be detested by anyone from any class.

14

不中矢ハ不放事

Atarazaru Ya ha Hanatazu no Koto

THE ART OF NOT SHOOTING AN ARROW IF YOU ARE GOING TO MISS

Lower tradition:
Unless you are sure of the method of infiltration, you should not infiltrate. It is not good to infiltrate without a fair prospect [of success] and to take a risk of all or nothing.

Middle tradition:
Choose those shinobi no mono who can make a suggestion of what technique should be used, as soon as you have talked to them [about the upcoming mission], and send them out. You should not use those who are slow on the uptake.

Upper tradition:
If you execute a plan and it fails, you should not try with the same method again. Try with another art[99] the next time.

15

カヘリ問ノ事

Kaeri toi no Koto

THE ART OF QUESTIONING [YOUR SPIES]

Lower tradition:
You should not carry out an enquiry [into the information gathered by shinobi] with several shinobi no mono together at one place. Instead you should listen to them one by one, so that you will find out if there are inconsistencies or not between them all. If there is something you suspect, you should question them again and again at intervals.[100]

Middle tradition:
Assuming you have questioned Gen'emon,[101] then you should question Yaemon about the points you got from the inquiry with Gen'emon. Then question Gen'emon on the points that have come up in the inquiry with Yaemon.

Upper tradition:
Make up something that is not included in either Gen'emon or Yaemon's report and ask them about this false subject, in this way you will discover if they are trustworthy or not. Furthermore, you should be fully aware of these techniques, whether you are using them or if others are using them on you.

This is the end of Master Yoshitaka's teachings from Koka.

NOTES

1 It is unknown which two specific scrolls he is referring to here.
2 Shinobu 忍 - in secret.
3 Jindai Jokan神代上巻however Chikamatsu is referring to either the document Kojiki or Nihonshoki.
4 That is, created by the gods but used by humans.
5 Thus Japan's spying arts are divinely inspired, whereas those of other nations are only the ideas of man.
6 The point here is that the pheasant has no name or is unnamed and therefore is not a noble or 'respectable' enough origin for the shinobi arts.
7 中士以下.
8 'Setto' or stealing and thievery, this set of ideograms is often used to mean shinobi in military texts of the Edo period.
9 Meaning all of the land of Japan has different manners and customs.
10 Most likely Iga.
11 After the Sengoku Period.
12 In a time of war have a trained military person undertake this but in peacetime take some-one from the lower levels of society, as it is less obvious because they are expected to move between areas.
13 This paragraph is additional information inserted from the Yokan Denkai Kudensho scroll.
14 Oribei 折塀 A form of wall in Japanese castles that is like folded screens and has a series of indents which form corners.
15 Additional information from the Yokan Denkai Kudensho scroll.
16 This point is repeated because this section is additional information from the Yokan Denkai Kudensho scroll.
17 Ban'no 万能 - universal tool, most likely a grappling hook.
18 The word 'consider' here is probably used because the shinobi cannot see up into the darkness of the underside of the roofing.
19 The *Go* insert is taken from the Yokan Denkai Kudensho scroll.
20 That is, do not be obvious by attacking from the opposite direction.
21 I.e., it is an illusion and has no substance.
22 This paragraph is additional information from the Yokan Denkai Kudensho scroll
23 Chikamatsu is saying that a doctor observes, listens, questions and treats a patient and it is the same in converting a spy.
24 This passage is additional information from the Yokan Denkai Kudensho scroll.
25 The inference here is that you 'sound out' the enemy and try to gain information from them by strategic means.
26 This sentence is additional information from the Yokan Denkai Kudensho scroll.
27 The section here is an amalgamation of both texts.
28 This sentence is additional information from the Yokan Denkai Kudensho scroll.
29 Additional information from the Yokan Denkai Kudensho scroll.
30 Additional information from the Yokan Denkai Kudensho scroll.
31 This sentence has been slightly changed to reflect better the meaning. The main point here is that the enemy tried to use a small defeat to get you into a position where they could gain a significant victory, however, because you can separate from the need to gain this small benefit you will win the overall contest.

32 Information acquired can create a 'Catch-22' situation.
33 Mawashi Zonae – a type of military formation where different troops attack at different times. A similar version appears in the Bansenshukai ninja manual.
34 This additional section is an extended explanation of the above from the Yokan Denkai Kudensho scroll.
35 Hatamoto.
36 This implies that he is lecturing to, or teaching, Naganuma Ryu at this point.
37 The text says in truth 真 and falseness 偽 – meaning the observable and physical and that which was the unseen game of tactics behind the reality.
38 矢モ楯モタマラセヌ強気
39 That is, remain strong in your mind, take time mentally and do not rush for an exit, as it will lead to failure.
40 So the shinobi should have a determined mindset, the dogs know the shinobi will kill them and it appears that the dogs 'pick up' on this and leave the shinobi alone.
41 This can be any form of fortification.
42 Shicho 紙帳 A paper cube hung from the celling to act as a net to keep insects at bay.
43 半弓, actually these bows, while shorter than a normal bow are still over six feet tall.
44 Shinobi no Souto 忍ノ總頭.
45 Additional information from the Yokan Denkai Kudensho scroll which ends at this point.
46 The information from the Yokan Denkai Kudensho starts again here.
47 This could be seen in two ways, either this is pre-planned and is about taking down enemy guards, or, if combat is entered, teams of two or three pick individual targets and fight as a unit in open combat.
48 If you wish to retreat and do not feel the opportunity is correct to attack the enemy, then you should first attack the outer guard with this method and declare that the fight will be unlucky and retreat. How to retreat from a situation you are not happy with by claiming that omens are bad.
49 Most likely a fire weapon on a pole.
50 In Japanese, Hiburi taimatsu no jutsu.
51 This number should be 25, however, Chikamatsu repeated the number 24 and thus all the rest of the articles are out of sync by one place in the original text – we have corrected them in this translation.
52 Shinobi no Ie 忍ノ家 this can be translated as family, or school.
53 The actual title of the play does not have the name Kumasaka in it, but he writes it two different ways, once with the name added and once without.
54 He is pointing out here the connection between shinobi idioms and the study of Sun Tzu.
55 A very similar idiom is used in Japanese.
56 This last paragraph is additional information from the Yokan Denkai Kudensho scroll.
57 This section is totally new and is wholly from the Yokan Denkai Kudensho scroll. It has not been numbered because it was an extra slip attached to the face of the page and is connected to the above section.
58 The length of the 'stabbing sword' is uncertain and may be just a knife, therefore it has been translated as blade.
59 Ryo 両
60 This should probably be Ryo and not Koku, however the word Koku (a unit for salary) appears twice, so while it appears far too high, it is not impossible)
61 One bu, a fourth of one ryo gold coin.
62 Masterless samurai.
63 Interestingly these two retainers were ordered to commit suicide by Tokugawa Ieyasu due to his suspicion about their loyalty to the eastern forces.
64 Additional information from the Yokan Denkai Kudensho scroll.
65 Additional information from the Yokan Denkai Kudensho scroll.
66 Men from Koka, meaning Koka ninja.
67 Possibly read as Jo wo Shiru no Daiji.
68 A manuscript written presumably in the Sengoku period about the character of the residents of each province. It is said Takeda Shingen was an enthusiast and used to add notes if he got additional information from his spies.

69 Literally 'Art of the Seven Ideograms'.
70 In the original text it literally says 'know yourself and think of this art' but his point here is to know how you and others work, based on these seven points.
71 Kimura Okunosuke of Koka.
72 All of these Koka skills are divided into three levels; sho 初, chu 中go 後 and have been translated here as lower, middle and upper. The idea is that a shinobi would start with the lower traditions and make his way up to the upper traditions, which are the most secret sections of the art.
73 The ideogram for liver also means 'courage' or to have 'nerve' 肝 and is thus repeatedly used in Koka traditions.
74 Most likely the upper and lower traditions have been switched by accident here as magical elements are normally considered the highest form.
75 Kusabagakure.
76 Presumably under your breath.
77 The Wisdom Sutra has 262 ideograms.
78 The Japanese used to be buried in barrels.
79 Again this appears to be the wrong way around as it is too simplistic and should be lower tradition.
80 Not to be confused with Fukushima Ryu 'Blind Buddha hiding'.
81 *Atypus karschi* – the Purse Spider which lives in a hole underground, hence the skill-set.
82 When making a fan, put two sheets of paper together, one on each side of the fan ribs and insert the powder between these two sheets, making it a magical fan.
83 By repeating the name of Amida Buddha in Japanese, 'amuamidabutsu' it is thought in Buddhism that one can obtain birth in Amida's pure land.
84 *Piper nigrum*.
85 *Zanthoxylum piperitum*.
86 That is, you should infiltrate under the cover of mist.
87 The small dot in the upper right corner is the last in the order of strokes for this ideogram. What he means here is to write the ideogram in the air, but as the dot is the last stroke of the ideogram, the action would be the shinobi pointing at the dog and this would happen three times as the ideogram should be written three times. Interestingly, this same ideogram appears in the ninja scroll *Mizukagami* where it is used to force a horse on to a ship, again using the last dot as a signalling motion.
88 This tradition is found in the Gunpo Jiyoshu manual and can be found in *The Secret Traditions of the Shinobi* – it is considered to be from the Hattori family.
89 Possibly fish bones but as yet unknown.
90 This appears in the Shoninki manual which we translated as 'Dignity of Flying Birds' due to the context in which it was found. However, the skill-set is totally different in this tradition.
91 Simply the feeling of jumping up again to try to soften the landing.
92 A polearm.
93 One shaku (30cm) is the length above the belt, which you hold on to.
94 Literal translation of the corrected poem: 'I am now wondering if you have forgotten to make a promise that you are not to betray, seeing a branch of the wisteria flower.'
95 This could be read as 'inform him'.
96 Once again, these traditions, lower and upper, would appear to be in the wrong order. The first half of the poem is not in standard Japanese, however the second half reads: 'I am going to sleep now, thus if something serious should happen, please surprise and wake me up, my god of the pillow.'
97 This poem is from Kokin Wakashu, 'A Collection of Ancient and Modern Japanese Poetry', which was an imperial anthology of poems from the tenth century.
98 This is actually the dried straw of the rice plant, and the leaves would be dry and the stem would be straw like.
99 Jutsu 術
100 The original line appears to cut off as though it should continue or else it ends extremely abruptly.
101 These are example names, (Peter, Paul).

THE SECOND PART OF THE SCROLL BY MASTER YORIHIDE OF IGA

蘊奥口授傳目

Un'o Kuju Denmoku

ORAL TRADITIONS ON THE DEEPEST SECRETS OF IGA

1

撰ム事

Eramu Koto

THE ART OF HOW TO CHOOSE [PEOPLE]

Some shinobi no mono are good at big skills 大業 while others are good at small skills 小業.[1] It is essential for the captain[2] to use them according to the correct judgement of his men's characters. You should not assign your men to do what they are not good at.

2

闇ノ夜ニ礫ウツヘカラサル事

Yami no Yo ni Tsubute Utsubekarazaru Koto

THE ART OF NOT CASTING STONES ON A PITCH BLACK NIGHT[3]

Do not attempt things where you are unsure of the success level and where you need to engage in them to know the results. Instead you should only commence your activities if they have good prospects at each step throughout your entire plan.

3

大小ノ心得アルヘキ事

Daisho no Kokoroe Arubeki Koto

THE ART OF JUDGING HOW MANY PEOPLE YOU SHOULD USE

You should have a good think about where and when a large number of people would be best used. Also, you should judge wisely if a small number will suffice or not – in short, do not waste the power of your forces.

4

多少ノ心得アルヘキ事

Tasho no Kokoroe Arubeki Koto

THE ART OF JUDGING HOW MUCH MONEY SHOULD BE USED

When you use gold or silver, you should think of how much money should be used. In the case that a small amount of money is better, if you spend too much, it may do harm, while in a case where a large amount should be used and if you spend too little, it will be even worse. This all depends on the matter at hand, the time and the people [involved] and every detail cannot be described here.

5

ツカラカシノ傳

Tsukarakashi no Tsutae

THE TRADITIONS OF EXHAUSTING THE ENEMY

If the target position is strictly secured and has tight defences in all directions and it is difficult for you to infiltrate, you should divide your force into several groups and have each group simulate false night raids in the middle of every night and in turns. The purpose of this is to make the enemy tense, as they think you will attack at any moment but actually, you do not. After two or three nights of doing this, every enemy solder will be tired and if they are tired, they will offer a gap which you should take advantage of and infiltrate.

6

現問ノ術

Utsutsudoi no Jutsu

THE SKILL OF INTERROGATION BY SLEEP DEPRIVATION[4]

If you have captured an enemy spy alive but he does not give way no matter how hard you question him – furthermore, if he is a bushi warrior, it is even less likely that he will give away information if you torture him – in such a case, use the following technique while interrogating him. You should divide the guards into groups and make them take turns when questioning and provoke him loudly, day and night, this is done so that the prisoner cannot sleep. If he cannot sleep for a few days and nights, he will be exhausted in his mind and fall into a condition called 'clouded consciousness' 気脱ケ. Then, at this point you should have the guards question him and get a confession in this manner.

7

火中状ノ傳

Kachujo no Tsutae

THE TRADITION OF THE LETTER IN THE FIRE

Seal a letter with gunpowder rolled within it. As soon as you put this letter into a fire, it will burst into flames immediately. If a normal letter is thrown into a fire it will take a while to destroy.

8

形ヲ隠ス術

Katachi wo Kakusu Jutsu

THE SKILL OF HIDING YOUR IDENTITY

This is a technique to hide your identity after you have killed yourself. You should put a substantial amount of gunpowder inside [the base of] your futon and make sure it is dry. When you kill yourself, you should set fires all around you, then lay out the futon.[5] You should kill yourself while sitting on the futon so that after you die the futon will ignite and fire will rage around your body. This way you will be burned to ashes and it will be hard to identify whose body it is. Also, it is most likely that even the bones will be shattered and dispersed.

9

切火縄ノ所作

Kiri Hinawa no Shosa

THE TRICK OF CUT FUSE[6]

When you retreat after your covert activity, you should light cut fuses and put them here and there, held in place by [split] bamboo sticks or branches. Meanwhile you should take a detour route far away [from these lights]. Those following you will aim at the fuses you have set up.

10

マカクレノ傳

Magakure no Tsutae

THE TRADITION OF HIDING FROM SIGHT

Magakure is sometimes taken to mean *a secret skill to hide at a certain distance* because the ideogram Ma 間 means [one] Ken in length, but this understanding is not correct. In truth, Magakure means 'hiding from the eyes', as Ma 目 here means 'eye', thus you hide yourself from enemy eyes.

One tradition says that you should dry Purse Web spiders in shade and powder them and keep the powder in a cylinder. You should carry this on your waist and when the enemy get close to you, scatter it over the enemy. The enemy will cough and be blinded with this powder. This skill is called Makakushi no jutsu[7] マカクシノ術.

However, as well as this, any other way you use to avoid detection by the enemy is called Makakure.

11

霧之印ノ大事

Kiri no In no Daiji

THE PRINCIPLE OF THE MUDRA OF MIST

Ingredients

- Sulphur – 1 ryo
- Saltpeter – ½ ryo
- Ground pepper – ½ ryo
- Blue vitriol – 1 ryo
- Chili pepper – ½ ryo
- Nicotine of tobacco to be dried – ½ ryo

Grind the above into fine powder and put it into a bamboo cylinder. Put the cylinder onto a spear shaft or on your waist and scatter it over the enemy when the enemy gets close to you.

Also, it can be used to capture those who are holed up in a building. Be sure that you take a windward position.

12

地蔵薬師ノ前ウシロノ心得

Jizo Yakushi no Mae Ushiro no Kokoroe

HINTS FOR BEFORE AND AFTER [THE FESTIVALS OF] YAKUSHI[8] AND JIZO[9]

In principle, a shinobi relies on darkness. The day of festival for Yakushi is the eighth day [of the lunar month]. Before the eighth day the moon appears in the early evening. The day of the festival for Jizo is on the 24th day [of the lunar month] and after that day you have moonless nights. Therefore, you should be careful [of infiltrators] before and after the festivals of Yakushi and Jizo. When you infiltrate, you should also keep this principle in mind.

13

足カラミノ術

Ashi Karami no Jutsu

THE SKILL OF THE FOOTHOLD TRAP

Where you think shinobi will sneak in from, fix brackets[10] and stretch rope across the area. The height of this should be varied. If the enemy stumble over the rope, they will become cautious, and if they become cautious they will not easily take steps forward. This art is not only about stretching rope but you should apply this principle to everything.

14

腕カラミノ術

Ude Karami no Jutsu

THE SKILL OF THE ARM-RESTRAINING TRAP

As well as with the above skill, there is the skill of leaving a sword scabbard and its cord on the route you expect shinobi will come from. If the enemy stumble over this scabbard and find it, it looks like the sword has just been drawn [and that the scabbard has been left behind], they will become worried and have to pay attention to all directions, thinking that you have drawn your sword and are watching for a chance to attack them. This prevents them from moving forward and they will be neutralised, so this skill is called 'The Arm-restraining Trap'. Besides the above example given, this skill of The Arm-restraining Trap should be varied with all the resourcefulness you can muster.

15

道ヲ道トスヘキ傳

Michi wo Michi to Subeki Tsutae

THE TRADITIONS WHEN WALKING MAIN HIGHWAYS

The first principle is to disguise yourself as someone who travels on main roads. If you do not travel along the main roads, it is difficult to use express messengers or other measures when in an emergency. On the boundaries of your own province, you should make arrangements for prepared horses so that an express message can be carried by a post-horse courier.

16

後途ノ勝ニ名ヲ残ス事

Godo no Kachi ni Na wo Nokosu Koto

THE ART OF LEAVING YOUR NAME FOR THE TIME OF VICTORY

When you return after completing a covert mission and if it is the case that it is acceptable to make the enemy realise that you have infiltrated far into their territory and that you are such a good shinobi – showing them how dangerous it is, as they have been so negligent to allow you to do that – then you may leave clear evidence that proves that you were there. It is called Na wo Nokosu 名ヲ残ス [The Art of] Leaving Your Name.

17

大勝ノ為ニ名ヲ惜ム事

Taisho no Tame ni Na wo Oshimu Koto

THE ART OF HIDING YOUR NAME TO AID A GREAT VICTORY

If you need to infiltrate again after you have already infiltrated successfully once, you should destroy all traces of your infiltration before you leave, this is done so that the enemy will not notice that you have been there. By doing this, you can infiltrate again with ease. There are things you should be aware of when you bring any form of evidence back [from your mission].

18

枕ヲトル事

Makura wo Toru Koto

THE ART OF TAKING THE PILLOW

According to one story, there was a man who took the pillow that an enemy general was using that night when the general was advancing in another's province. After this a messenger was sent to give the pillow back to him with a message saying, 'One of our retainers happened to find this pillow last night in front of your position[11] and so we would like to return it to you.' The enemy general was shocked and thought that as they had such an excellent shinobi, it would be easy for this agent to assassinate him while he was sleeping. Because of this he then immediately pulled his army back to his own province.[12]

Someone commented on the story and this was what they said:

> Rather than stealing the pillow, he should have killed the general with one stroke of the sword. [He then said] the agent could not have actually killed him because [shinobi skills] are based on the magic of monsters[13] and therefore [shinobi] skills are not substantial. The enemy general could have answered this message in a proper way [as opposed to fleeing] and it was regrettable that he was filled with fear.

You can tell that this comment is completely ignorant and that the commentator does not know anything about shinobi no jutsu [and he did not know that] this feat must have been achieved by the use of Converted or Internal Spies. [What would have happened in reality is that] you would have told the spy planted within the enemy to take the pillow and that if he succeeds he will be rewarded with tens of thousands of gold coins. That spy would then take the pillow and sell it to your side, but he would not go as far as taking his lord's head. This is why he could not have taken the lord's head but only took his pillow. Furthermore, this theft is a serious crime, so the spy will keep it secret and this secret will not be discovered for a long time. Take note; in case you need to use the same technique in the future, you should not even tell your lord that you have used this skill of 'taking the pillow' so that this skill will remain unknown. Thus, it will turn out that the story will be handed down that a shinobi no mono dared to infiltrate as deep as the bedroom of the enemy and bring only a pillow. You should notice that Converted Spies or Inward Spies can be useful in this way as well.

19

毒薬反シテ薬トナル術

Dokuyaku Hanshite Kusuri tonaru Jutsu

THE SKILL OF POISON WHICH WILL IN TURN BECOME MEDICINE

Although it is an unrighteous and morally wrong way to kill someone with poison, it can be righteous if the purpose is to destroy the unrighteous with justice. This is because you can fulfil your aim without a massive loss of life. Thus, as a result, by poisoning only the enemy general, it will in fact work as a form of medicine for the victory of your army as it allows you to be victorious without fighting.

In this skill you should use your Internal Spies, so that they will have someone within the enemy carry out [the poisoning]. This should be done by bribing the person with a large amount of money. Remember Sun Tzu warns that you should not withhold rank or salary [from your spies]. It is easier to win a war without drenching your blades with blood by the use of money. If you make an ally of someone within the enemy ranks and have that person conduct [the assassination], it will not be discovered for some time and in this way you achieve a complete victory.

In earlier times, there were two brothers who were generals in a province. The younger brother intended to ruin his brother but things were peaceful and it was difficult for him to take up arms against his kin. So he tried various kinds of magic and spells but they failed. Meanwhile, the traitorous brother had a favourite retainer who had been serving him since he was a child, and this brother conspired with this retainer, who was so greedy and cruel that he was very happy to form a plot with his lord. This was wrong of him and this retainer should have protested against his lord's plan, even at the cost of his own life.

The older brother had a doctor who was new and favoured because he talked smoothly and had a way with flattery. Thinking that this doctor afforded the ideal opportunity to support the plot, the younger brother's retainer approached the doctor and won him over with the promise of gold when the aim was accomplished. In time the older brother succumbed to a disease and the doctor was given the opportunity to give him medicine. Taking advantage of this opportunity, he gave the lord poison and killed him. The younger brother successfully took over the family and gave his favourite retainer a huge amount of money and a very high rank. The doctor left [the clan] because the lord died; however, the younger brother favoured him and rehired him as a close retainer as well as giving him a massive amount of money as promised.[14] The doctor became very well off and extremely arrogant. However, the situation reflected a popular saying: 'Heaven knows, Earth knows and so people will know'[15] and rumour spread widely but nothing came of this suspicion.

This stratagem should have been flawless and should not have been leaked, so why is it that their plot came to be known? Well, it is because of a servant under the doctor. This servant came to a neighbour's house and complained that his master had massive amounts of money but he was such a miser [and as the servant saw] the doctor counting his money he concluded there was as much as 13,000 gold coins in his horde. Another servant said that his master [the doctor] kept a chest in storage which he referred to as 'treasure'. When the doctor visited Edo, he left this chest for safekeeping with someone and had it back on his return. However, when the chest was carried, it took four strong men and it was not easy even for them as a group to carry it. Someone who heard this said: considering how much money he earned for his medical practice and fief – even if his whole working life was taken into account – and how much money he could have saved or gained through profit and also thinking of the cost of the style in which he lived, how his children were faring, how much money he had spent on giving presents to curry favour with various people, he should not have had more than about 3,000 pieces of gold. Therefore, it was suspicious that he had such a massive amount of money. And so it was decided that it must be the reward for the poisoning, as was rumoured at the time. [Remember, how things are detected and found out] can be absolutely unbelievable. This episode is written here as a warning for all future students.

20

辞ニ花ヲ咲セル辨

Ji ni Hana wo Sakaseru Ben

THE ART OF MAKING A CONVERSATION BLOSSOM

To invite and encourage someone to be a Converted Spy, Internal Spy or traitor, you should enter into the person's mind through 'blossoming' conversation to produce the 'fruit' of your labours.[16] The basic ways are thus:

1. Tell him what offences the enemy have committed
2. Of justice and injustice
3. Which side will be more beneficial
4. Tell him you are offering a lot of money in return for his service.

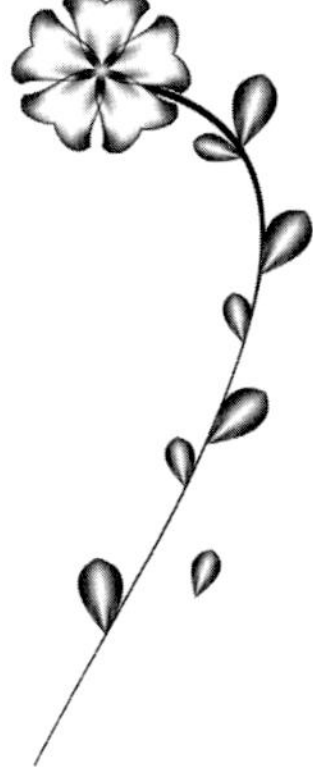

Other than these, you should work out various ways to achieve [the 'blossoming' of the conversation] according to the situation. If the person is not moved by discussing justice and injustice, he will be influenced by reward, while if he is not moved by money, he will be by the sense of justice. If not by either of these two, you may have to overwhelm him by the power of your side's military capability.[17]

21

チラシ書ノ傳

Chirashi Gaki no Tsutae

THE TRADITIONS OF WRITING A SERIES OF FAKE LETTERS

You should write various fake letters and drop them by Yabumi, which are letters sent by arrows or by having shinobi leave them in the enemy compound or having them found by an enemy spy. Basically, this skill should be carried out after you have employed the *Art of Three-way Connection* or the *Art of Secret Connection*. One letter should say such things as, 'We are glad to hear you are going to betray them.' Also it should reveal the ways of sending signals or confirming the promise of the reward 'he' will be given. Another letter should say, 'Attack them from the rear and set fires, and you will be rewarded.' Another letter should give no clear explanations [of any form of military action] but instead should give something that looks like a password or identifying marks and will make the enemy suspicious about who the intended receiver is, or make them wonder as to the deeper meaning of the letters [and the implications]. Another letter should ask for others to rise in rebellion from within the enemy side. You should arrange tens of such letters that will get into enemy hands and then send shinobi to disperse such information. By these measures, the enemy will be suspicious of each other and conflicts will arise.

22

大事ノ使ノ術

Daiji no Tsukai no Jutsu

THE SKILL OF SENDING AN IMPORTANT MESSAGE

To send important messages you should make all arrangements beforehand with each other. For example, if the date is the seventh day of the seventh month, you should send seven pieces of fish[18] each of 7 sun in length. To promise to [carry out] treachery, you should send salted fish. When you are going to commit arson, you should send dried fish. When you are asking for provisions, you should send rice cakes. When you ask for reinforcements, you should send sweet cakes, when asking for forces to attack the enemy from the rear, use rolls or something of that ilk. When you send these things, you should attach a normal letter [which will not draw suspicion], but the receiver will understand the message as you have arranged these things beforehand. In the case where the messenger gets captured by the enemy, he does not know anything if you use this method and so he has nothing to confess. These items are what you usually send to others, so that people will not notice anything unnatural, and if they do, they cannot figure out the hidden meaning.

23

声ヲ知ルヘノ傳

Koe wo Shirube no Tsutae

THE TRADITIONS OF USING VOICE AS GUIDE

When you have infiltrated on a pitch dark night and if you are detected and have to fight, you should not utter a word. Normally [in the dark] you have to attack each other by listening to any sounds from the enemy's mouth.[19] Therefore, if you find that a spy has infiltrated your position, you should be careful of any sounds you make which may come from your mouth. This is because the enemy will attack you by using any sound you make as a guide to find you. Without any sounds made, the enemy will be vexed.

24

大スワリノ傳

Dai Suwari no Tsutae

THE TRADITION OF GREAT STABILITY

This point is one that any leader should keep in mind. If you are using shinobi by sending them in various directions and in the case [where the plan or shinobi] is detected, you should secure a method to allow the primary aim of the plan to remain a possibility.[20] How to do this depends on the person, matter or situation so details are not transmitted here. Know that it is wrong to send spies without using this art.

This is the end of Master Yorihide of Iga's teachings.

NOTES

1 Most likely, that some shinobi are good at overall strategy while others have specific skills. However, the text does not explain in more detail.
2 Soushi 總司
3 Guesswork and estimation without information.
4 The original title is more poetic and concerns itself with the time between sleep and consciousness.
5 Take it out of its storage and make it as if you were preparing for sleep.
6 Most fuses are in coils. The cut fuse is a short length of fuse taken from the roll.
7 While different terms, both this one and the above are almost identical in application.
8 The healing Buddha.
9 The guardian deity of travellers and children.
10 This could be to fix or to place next to a wall, therefore it could be rests or brackets.
11 This obvious lie is intended to intimidate.
12 This story also appears in a similar form in Chinese literature, see *In Search of the Ninja*, page 69.
13 Yokai 妖怪.
14 This sentence is ambiguous and partially reconstructed.
15 No matter how deeply you try to keep something secret, heaven and earth (ie the supernatural) always know what you have done and therefore the secret will come out in the end.
16 The analogy is one of sowing and reaping to achieve an aim.
17 This is ambiguous, while it sounds like you should impress him with military force, it could be a direct threat.
18 肴 can be read as fish or small sections of food.
19 The text says 'voice' but context implies breathing, grunts, war cries, etc.
20 That is, there should be a backup for the plan even if it has been discovered that will achieve the same aim.

忍法行巻

Shinobi[1] Hogyo no Maki

THE SHINOBI WAY OF DIVINATION SCROLL

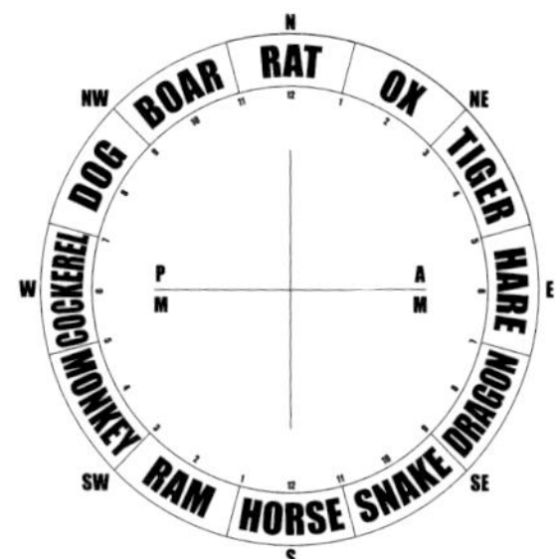

年王位之事

Toshi no Oi no Koto

THE ART OF KNOWING WHICH LUNAR MANSION IS PROMINENT IN WHICH YEAR

- The Year of Rat – the Maid mansion
- The Year of Ox – the Southern Dipper mansion
- The Year of Tiger – the Tail mansion
- The Year of Hare – the Chamber mansion
- The Year of Dragon – the Neck mansion
- The Year of Snake – the Wing mansion
- The Year of Horse – the Star mansion
- The Year of Ram – the Ghost mansion
- The Year of Monkey – the Three Stars mansion
- The Year of Cockerel – the Mane mansion
- The Year of Dog – the Bond mansion
- The Year of Boar – the Encampment mansion

In each of the above years, the lunar mansion connected to it is the most important for that year.

支干王主之事

Shikan Oshu no Koto

THE ART OF THE RELATIONSHIP BETWEEN A GIVEN ELEMENT IN ASSOCIATION WITH THE TWELVE SIGNS OF THE CHINESE ZODIAC AND THE TEN CELESTIAL STEMS

- Tiger, Hare – Wood
- Snake, Horse – Fire
- Monkey, Cockerel – Metal
- Boar, Rat – Water
- Ox, Ram, Dragon, Dog – Earth

[Also]

- Kinoe, Kinoto – Wood
- Hinoe, Hinoto – Fire
- Tsuchinoe, Tsuchinioto – Earth
- Kanoe, Kanoto – Metal
- Mitsunoe, Mizunoto – Water

四季王相之事

Shiki Oso no Koto

CONCERNING THE SEASONS AND THE CHANGING POWER OF FIVE PHASES

These five ideograms are called Goi, the Five Phases.

1. 王 (flourish)
2. 相 (aid)
3. 死 (death)
4. 囚 (prison)
5. 老 (aged)

- 王(flourish) and 相 (aid) are strong phases.
- 死 (death) it is extremely weak.
- The phase of 囚 (prison) is weak but still has some power.
- In the phase of 老 (aged) means that any strength remaining has been sapped.

You should be concerned with which stage or phase matches which season. For example, for the three months of spring, *east* is matched with 王 (to flourish) [when you study the lists below] therefore, the flourishing or strong direction for those times is east. You should consider this accordingly.

For the three months of summer, *south* is the direction of strength and intercardinal directions are matched to the 相 (aid) phase. West is connected to 死 (death), north is connected to 囚 (prison) and east is connected to 老 (aged) [as seen in the list]. You should understand this and keep it in mind.

For each [of the four] Doyo periods in the year[2] the intercardinal directions are connected to the phase 王 (to flourish).

The intercardinal directions are:

- Ox is north-east-north
- Ram is south-east-south
- Dragon is east-south-east
- Dog is west-north-west

In spring the connections are:

- 王 – east,
- 相 – south
- 死 – intercardinal points
- 囚 – west
- 老 – north

The time of sunrise [is strongest in spring]
The hours of Tiger and Hare [are strongest in spring]

In summer the connections are:

- 王 – south
- 相 – intercardinal points
- 死 – west
- 囚 – north
- 老 – east

Daytime [is the strongest in summer]
The hours of Snake and Horse [are the strongest in summer]

In the period of Doyo the connections are:

- 王 – intercardinal points
- 相 – west
- 死 – north
- 囚 – east
- 老 – south

The hours of Ox, Ram, Dragon and Dog [are strongest in Doyo]

In autumn the connections are:

- 王 – west
- 相 – north
- 死 – east
- 囚 – south
- 老 – intercardinal points

Sunset [is the strongest in autumn]
The hours of Monkey and Cockerel [are strongest in autumn]

In winter the connections are:

- 王 – north
- 相 – east
- 死 – south
- 囚 – intercardinal points
- 老 – west

Night time [is strongest in winter]
The hours of Boar and Rat [are strongest in winter]

Apply the above information [to see the seasonal changes]. The time mentioned below each section is the most powerful for that time of the year.

月王座之事

Tsuki no Oza no Koto

THE MOST IMPORTANT LUNAR MANSION CONNECTED TO EACH MONTH

- The first month – the Encampment mansion
- The second month – the Stride mansion
- The third month – the Stomach mansion
- The fourth month – the Net mansion
- The fifth month – the Three Stars mansion
- The sixth month – the Ghost mansion
- The seventh month – the Extended Net mansion
- The eighth month – the Horn mansion
- The ninth month – the Base mansion
- The tenth month – the Heart mansion
- The eleventh month – the Southern Dipper mansion
- The twelfth month – the Emptiness mansion

The above applies for each and every year, such as; the Encampment mansion and its connection to the first lunar month of each year, the Stride mansion for the second lunar month, and the Stomach mansion for the third lunar month. The rest follows on as above.

日王分之事

Hi no Obun no Koto

THE ART OF KNOWING THE LUNAR MANSION BY THE FIRST DAY OF THE MONTH

- The first day of the first month – the Encampment mansion
- The first day of the second month – the Stride mansion
- The first day of the third month – the Stomach mansion
- The first day of the fourth month – the Net mansion
- The first day of the fifth month – the Three Stars mansion
- The first day of the sixth month – the Ghost mansion
- The first day of the seventh month – the Extended Net mansion
- The first day of the eighth month – the Horn mansion
- The first day of the ninth month – the Base mansion
- The first day of the tenth month – the Heart mansion
- The first day of the eleventh month – the Southern Dipper mansion
- The first day of the twelfth month – the Emptiness mansion

As shown above, the first day of each month starts with each of the above lunar mansions, so you should count up to the date you want to know. For example, if you wish to know which mansion the moon is in on the fifth day of the first lunar month, as the first day is the Encampment mansion, then you should count five from the Encampment mansion. Next would be; the Wall, the Stride, the Bond and then the fifth would be the Stomach mansion, making the Stomach mansion the mansion connected with that day. For the rest, the same should apply.

時王順之事

Toki no Ojun no Koto

THE ART OF KNOWING WHICH LUNAR MANSION IS MOST IMPORTANT BY THE HOUR

As was done above, apply the above lunar mansions for the hours of the day by starting with the hour of Ox (1-3am) – but you should skip the hour of Horse[3] (11am-1pm) [to find the mansion for that hour]. However, do not include the Ox lunar mansion.

相尅相生事

Sokoku Sosho [no] Koto

THE ART OF DESTRUCTION AND CREATION CYCLES

[and]

相性相尅事

Sosho Sokoku [no] Koto

THE ART OF CREATION AND DESTRUCTION CYCLES

The Creation Cycle

- Wood produces Fire.
- Fire produces Earth.
- Earth produces Metal.
- Metal produces Water.
- Water produces Wood.

The Destruction Cycle

- Wood destroys Earth.
- Earth destroys Water.
- Water destroys Fire.
- Fire destroys Metal.
- Metal destroys Wood.

Traditions say:
Creation Cycles represent the beginning of something while Destruction Cycles represent the end. At the beginning of all, the 'great source' 大極 came into being and heaven produced [the first drops of] water, then water produced wood, wood produced fire, fire produced earth, earth produced metal and metal produced water.

There are two ways to view the Creation Cycle; first is the *Taisosho* 'Objective Creation Cycle', and second is *Jusosho* 'Subjective Creation Cycle'. In Taisosho, you affect an external thing, and this is negative, while in Jusosho it is you that is affected, which is a positive thing.

Next is the Destruction Cycle, water destroys fire, fire destroys metal, etc. There are two kinds of Destruction Cycle, *Tai Sokoku* which is an 'Objective Destruction Cycle', where you affect something, which is positive and also there is the *Ju Sokoku* 'Subjective Destruction Cycle', where the object affects you, this is a negative cycle.

If any given situation that you are in has two elements which are the same, ie, wood to wood, water to water, fire to fire, earth to earth and metal to metal, then this is called Soka.

[In Soka] there are also two kinds; *creative* and *destructive*. If wood is added to wood, [that is trees are put together] then it makes a forest, if it is water to water, then it creates a flood, so they are of the creative types of Soka. If it is earth to earth, it makes mountains and this is also of the creative type. On the other hand, if fire meets fire, it makes flames, and if metal meets metal, they clash and scrape on each other, so they are of the destructive type of Soka.

Apply the cycles to the elements of Creation, Destruction and Soka (when two elements double up) when you perform something based on astrology.

If water is added to water, it is half-lucky. More is to be orally transmitted.

時取之事

Tokidori no Koto

THE ART OF CHOOSING A POSITIVE HOUR

On the day of Kinoe or Tsuchinoto, they are the Hours of Rat, Ox, Tiger, Hare and Dragon

On the day of Kinoto or Kanoe, they are the the Hours of Monkey, Cockerel, Dog, Boar and Rat

On the day of Hinoe or Kanoto, they are the Hours of Dragon, Snake, Horse, Ram and Monkey

On the day of Hinoto or Mizunoe, they are the Hours of Horse, Ram, Monkey and Cockerel

On the day of Tsuchinoe or Mizunoto, they are the Hours of Tiger, Boar and Dragon

方角変化之事

Hogaku Henka no Koto

THE ART OF CHANGING DIRECTION

	(The direction of life)	**(The direction of death)**
The first month	Rat	Horse
The second month	Ox	Ram
The third month	Tiger	Monkey
The fourth month	Hare	Cockerel
The fifth month	Dragon	Dog
The sixth month	Snake	Boar
The seventh month	Horse	Rat
The eighth month	Ram	Ox
The ninth month	Monkey	Tiger
The tenth month	Cockerel	Hare
The eleventh month	Dog	Dragon
The twelfth month	Boar	Snake

This is also called the 'directions of life and death', or the 'changes of direction'.

日出入之事

Hi no Deiri no Koto

THE ART OF LEAVING AND ENTERING

When you need to venture out in the Direction of Death [as listed above] and if your aim cannot be fulfilled, then you should first walk seven steps in the Direction of Life, [as listed above] then go across to the Direction of Death and [carry on in that direction]. You should write just the outline for the ideogram for Oni 鬼 on your left palm and chant the following spell twenty-one times before you set off:

> 日メクリヲ頼ムタノマレ行程ニ猶コノマシキ日メクリノ神
> *Himekuri wo Tanomu Tanomare Ikuhodo ni Nao Konomashiki Himekuri no Kami*

Then you can be protected from evil luck from bad directions.
Also, if on a ship, you should trace the outline of the ideogram for soil 土 [on your left palm].

河伯大水神之事

Kahaku Omizu Gami no Koto

THE ART OF THE GODS, OF RIVERS AND OF FLOODS

- For Boar, Hare, Ram – East
- For Monkey, Rat, Dragon – North
- For Snake, Cockerel, Ox – West
- For Tiger, Horse, Dog – South

These above directions are drastically unlucky. You should avoid them on your departure [if the direction you] set sail [matches the above list].

厭日之事

En'nichi no Koto

CONCERNING DAYS TO BE AVOIDED

- Kinoe – Dog
- Tsuchinoe – Horse
- Mizunoe – Dog
- Kinoe – Tiger
- Tsuchinoto - Boar

These above are very unlucky days [and should be avoided]. They are also called *The Days of the Death of the Five Dragon King.*

厭對日

Entainichi

THE OPPOSITE TO THE ABOVE UNLUCKY DAYS

- Kinoe – Rat
- Hinoe – Rat
- Hinoe – Dragon
- Mizunoe – Rat
- Mizunoe – Horse

These above are great and lucky days. They are also called *The Days of Life of the Five Dragon King.*

破軍星之事

Hagunsei no Koto

CONCERNING THE STAR OF HAGUN[4]

- The first month – the fifth sign [counting clockwise]
- The second month – the sixth sign [counting clockwise]
- The third month – the seventh sign [counting clockwise]
- The fourth month – the eighth sign [counting clockwise]
- The fifth month – the ninth sign [counting clockwise]
- The sixth month – the tenth sign [counting clockwise]
- The seventh month – the eleventh sign [counting clockwise]
- The eighth month – the twelfth sign [counting clockwise]
- The ninth month – the first sign [counting clockwise]
- The tenth month –the second sign [counting clockwise]
- The eleventh month – the third sign [counting clockwise]
- The twelfth month – the fourth sign [counting clockwise]

For example, if it is the first lunar month at the Hour of Rat, you should not go in the direction of Dragon, which is the fifth sign [counting clockwise] from the Rat. An easier way to find if this is the unlucky direction is to count back four spaces [anti-clockwise] from the sign of the hour you are in. Take note that the system changes in the above list in the ninth month. Notice that in the ninth month, if it is the Hour of Rat, then the Direction of Rat is unlucky. Remember; do not go in the above directions.

指神方之事

Sashigami no Kata no Koto

CONCERNING THE DIRECTION OF 'SASHIGAMI'[5]

- On the day of Rat it is the direction of the Dragon
- On the day of Ox it is the direction of the Cockerel
- On the day of Tiger it is the direction of the Boar
- On the day of Hare it is the direction of the Ram
- On the day of Dragon it is the direction of the Monkey
- On the day of Snake it is the direction of the Dog
- On the day of Horse it is the direction of the Ox
- On the day of Ram it is the direction of the Rat
- On the day of Monkey it is the direction of the Hare
- On the day of Cockerel it is the direction of the Horse
- On the day of Dog it is the direction of the Tiger
- On the day of Boar it is the direction of the Snake

The above directions are unlucky for everything and you cannot be successful in anything where these are concerned. This is also called the direction of Hoshojin 方障神.

天宮神之事

Tengushin no Koto

CONCERNING TENGUSHIN[6]

- In the years of Monkey, Rat or Dragon – the direction of the Boar
- In the years of Boar, Hare or Ram – the direction of the Tiger
- In the years of Tiger, Horse or Dog – the direction of the Snake
- In the years of Snake, Cockerel or Ox – the direction of the Monkey

If you go to the above direction, you will have good luck. If you go to the opposite direction given, having the lucky direction to your back, then both people and horses will die.

帰亡日之事

Kimonichi no Koto

往亡日之事

Omonichi no Koto

Concerning Kimonichi, which are Unlucky Days for Returning from a Mission and Omonichi, which are Unlucky Days for Venturing Out on a Mission.

	(Kimonichi)	(Omonichi)
• The first month	Ox	the seventh day
• The second month	Tiger	the fourteenth day
• The third month	Tiger	the twenty first day
• The fourth month	Ox	the eighth day
• The fifth month	Tiger	the sixteenth day
• The sixth month	Rat	the twenty-fourth day
• The seventh month	Ox	the ninth day
• The eighth month	Tiger	the eighteenth day
• The ninth month	Tiger	the twenty-seventh day
• The tenth month	Ox	the tenth day
• The eleventh month	Tiger	the twentieth day
• The twelfth month	Rat	the thirtieth day

忌遠行事

Toku e Iku wo Imu Koto

CONCERNING UNLUCKY DAYS FOR TRAVELLING

• The first month	The Days of Tiger
• The second month	The Days of Hare
• The third month	The Days of Dragon
• The fourth month	The Days of Snake
• The fifth month	The Days of Horse
• The sixth month	The Days of Ram
• The seventh month	The Days of Monkey
• The eighth month	The Days of Cockerel
• The ninth month	The Days of Dog
• The tenth month	The Days of Boar
• The eleventh month	The Days of Rat
• The twelfth month	The Days of Ox

The above are terribly unlucky days for travelling.

忌夜行事

Yoru Iku wo Imu Koto

CONCERNING UNLUCKY DAYS FOR TRAVELLING AT NIGHT

- The first, fifth and ninth months, look to the Days of the Ox
- The second, sixth and tenth months, look to the Days of the Hare
- The third, seventh and eleventh months, look to the Days of the Monkey
- The fourth, eighth and twelfth months, look to the Days of the Cockerel

The above are very unlucky days.

道虚日之事

Dokonichi no Koto

CONCERNING THE DAYS OF DOKO[7]

This is also called *The Tradition of Kokyo.*[8] You should head for this direction for anything which relates to winning or losing. For a number [of troops] above ten thousand in number, you should use the *Year of Kokyo*, for a number above one thousand, then use the *Month of Kokyo*, and for the number above one hundred, use the *Day of Kokyo*. In our school we exclusively use the *Day of Kokyo* and have called it the *Days of Doko* since older times.

The tradition is thus:
The Kokyo for a year:
In the Year of Rat, the Directions of the Dog and the Boar are of 'ko' while those of Dragon and Snake are of 'kyo'.

The Kokyo for months:
If the first month of the year is connected to Tiger, then the eleventh and twelfth direction clockwise is Rat and Ox, these are 'ko' and [if you count seven spaces clockwise] then the Horse and Ram are 'kyo'.

The Kokyo for days:
On a day of Horse, [count eleven or twelve spaces clockwise] the directions of Dragon and Snake are of 'ko' and those of Dog and Boar are of 'kyo'.

You should apply the same method for the Kokyo for the hour. But remember, in all cases [be it month, day, hour], the eleventh and twelfth directions from the sign in question are of 'ko'. To find 'kyo' in a simple way, remember that as it is six spaces out of twelve, all you have to do is look at the sign directly opposite the one for 'ko' to find it with ease.

天一神之事

Ten'ichijin[9] no Koto

CONCERNING TEN'ICHIJIN

[Directions that you should avoid because the god dwells there]

- For six days from the Day of Tsuchinoto-Cockerel, (the god) goes north-east riding on a snake and stays there.
- For five days from the Day of Kinoto-Hare, (the god) goes east riding on a fish and stays there.
- For six days from the Day of Kanoe-Monkey, (the god) goes south-east riding on a hawk and stays there.

- For five days from the Day of Hinoe-Tiger, (the god) goes south riding on a cockerel and stays there.
- For six days from the Day of Kanoto-Ram, (the god) goes south-west on a deer and stays there.
- For five days from the Day of Hinoto-Ox, (the god) goes west on a horse and stays there.
- For six days from the Day of Mizunoe-Horse, (the god) goes north-west on a boar and stays there.
- For five days from the Day of Tsucninoe-Rat, (the god) goes north on a turtle and stays there.
- For sixteen days from the Days of Mizunoe-snake, (the god) stays up in heaven therefore, north until the Day of Hinoto-Cockerel.[10]
- From the Day of Tsuchinoe-Dog to the day of Mizunoto, (the god) stays in the west.
- From the Day of Kinoe-Dragon to Tsuchinoto-Monkey, (the god) stays in the east.
- On the Day of Mizunoto-Hare, (the god) stays in the south all day.

You should not fight facing the above directions.

箭放日事

Yahanachibi no Koto

CONCERNING THE DAYS WHEN YOU SHOULD SHOOT ARROWS [TO DISPEL THE NEGATIVE LUCK IN THE DIRECTION OF TEN'ICHIJIN]

If you cannot help but fight looking toward the direction of Ten'ichijin or fight on an unlucky day or in an unlucky direction, on a lucky day shoot two Kaburaya arrows[11] in a lucky direction for the enemy, to change the enemy's luck. By doing this you can also dispel the bad luck of that day or direction.

The days:

- In the first and seventh months, look to the Days of the Rat and the Horse
- In the second and eighth months, look to the Days of the Ox and the Ram
- In the third and ninth months, look to the Days of the Tiger and the Monkey
- In the fourth and tenth months, look to the Days of the Hare and the Cockerel
- In the fifth and eleventh months, look to the Days of the Dragon and the Dog
- In the sixth and twelfth months, look to the days of the Snake and the Tiger

The above days are lucky for all things.

The directions:

- On the Days of the Rat, Horse and Hare, it is the ninth direction (clockwise)
- On the Days of the Ox, Ram, Dragon and Dog, it is the same direction (as the sign of that day)
- On the days of the Tiger, Monkey, Snake and Boar, it will be the fifth direction

These directions are of the God of War, so you should shoot with these directions at your back.

三宿之事

Mishuku[12] no Koto

CONCERNING THREE MANSIONS

The months:

- The first, fourth, seventh and tenth months – the direction will be the twelfth animal (counting clockwise) including the animal of the day.
- The second, fifth, eighth and eleventh month – the direction will be the eighth animal (counting clockwise).
- The third, sixth, ninth and twelfth month – the direction will be the fourth animal (counting clockwise)

The above directions are beneficial for you if your shinobi or night attacks approach from these directions. Traditions state that these directions are called *Marishiten no Ake no Kata* 摩利支天ノ明ノ方.

陰陽俱錯事

In Yo Tomoni Majiwaru Koto

CONCERNING IN AND YO INTEGRATED

In and *Yo*, good and evil, are all integrated with each other and it is difficult to separate them. Divination for a date, time or direction is also the same [that is they are combined together, within each other]. Therefore, you should have good judgement about that which you should adopt and that which you should ignore, appropriately and according to the situation. Also, you should keep your mind balanced and calm. It is essential to decide what you should use and what you should not.

More is to be orally transmitted.

NOTES

1 The front cover of the original manual does not list the term 'shinobi' but it is used at the start of the manual itself, so is inserted here.
2 Doyo is the eighteen-day period before each new season begins, that is, the last eighteen days of spring before summer.
3 The text here is highly ambiguous.
4 The word *Hagun* 破軍 came from the Chinese name for one of the stars of Ursa Major. The star is the seventh star of the constellation and is the end star of the 'handle' of the 'dipper' shape. *Hagun* 破軍, literally means: 'destroying army'. In *Onmyodo* magic the direction that this star points was considered very unlucky and if you fought facing this direction you would be defeated.
5 Literally, pointing or finger god.
6 Literally, heaven palace god.
7 Literally, days of the unguarded road.
8 'ko' represents an animal Zodiac, to find which animal is 'ko' count clockwise and it will be the seventh and the twelfth animal in the system. To find the section 'kyo' you have to count clockwise to the seventh Zodiac animal, this will be 'kyo'.
9 Ten'ichijin is one of the twelve deities known as the Twelve Heavenly Generals in Onmyodo magic.
10 This referance to north may be a new sentence. The orginal manual splits the text into two here.
11 A screaming or whistling arrow.
12 Mishuku literally means three mansions.

法意巻

Hoi no Maki

THE VOLUME OF PRINCIPLES

陽幕之事

Yo Maku no Koto

THE ART OF YO WAR CURTAINS

[And]

陰幕之事

In Maku no Koto

THE ART OF IN WAR CURTAINS

Traditions say that it has been transmitted among tacticians that you should make two versions of war curtains to use; they are called Yo Maku and In Maku.[1] These screens are used to cover those places which you do not want to be seen. The first version is called In maku which covers things so that you cannot see them from the outside. People cannot see the inside from the outside but at the same time, the outside cannot be seen from the inside either. Therefore, the screens should have holes in them called 'monomi' – 'viewing things'.[2] Through these holes, those inside can see outside and those outside can see inside and for this reason, these screens with holes which allow you to see are called Yo Maku. Consider the above in full.

You will assume that you cannot see anything where these war curtains are spread all over, but remember that you can see through these Monomi viewing holes. As with these war curtains, no matter how strictly a place is guarded and even if there does not seem as if there is any gap or opening, there must be somewhere you can get in through a defence, just as these curtains have holes to see through. Tradition teaches you how to find such gaps.

Question: If screens have no monomi holes but are just sewed up without any openings at all, you cannot see or detect [what is on the inside], what should you do in that case?

Answer: If you cannot see inside then know that the opponent can equally not see you. However, if they do in fact have sewn-up screens with no holes, then they will be secure in the knowledge that no one can see through and thus they will let their guard down. This will allow you to take advantage of any gap and investigate with ease. Remember, sometimes there are chances that are presented that are outside of human knowledge and power.[3] Even if the curtains create a full cover, if a wild storm blows and the wind engulfs the curtains and they bunch together in the wind, then it will allow anyone to see inside or to even enter, not only on foot but even on horseback.[4] In turn, you should be very flexible when infiltrating by taking advantage of any change in circumstances. So remember, when it is difficult for you to infiltrate, you should wait for any change of circumstances [that can be taken advantage of]. Remember, the wind blowing upon the war curtains cannot be created by human power but is a natural phenomenon. When a situation is beyond your strength or wisdom, you should take advantage of the wind, rain, darkness, or other natural occurrences to fulfil your aim. You should be fully aware that war curtains are only one example of this principle [and that this principle extends to all things]. There are lots more oral traditions on this subject.

陽忍之事

Yo [no] Shinobi no Koto

陰忍之事

In [no] Shinobi no Koto

YO NO SHINOBI AND IN NO SHINOBI
SHINOBI OF THE LIGHT AND SHINOBI OF THE DARK[5]

There are three-fold traditions on this matter and they are as follows:
Koka traditions state[6] –

Lower tradition:[7]
Infiltrating during daytime is called Yo no Shinobi 陽ノ忍 while infiltrating at night time is called In no Shinobi 陰ノ忍.

Middle tradition:
If it is officially known to people who are hired as shinobi or as Koka-mono[8] and Iga-mono,[9] this is called Yo no shinobi. Kimura Okunosuke is an example of this.[10] On the other hand,

if your agent [has a cover] and has a different role [within your forces] but is undertaking the job of shinobi in a discreet way, or even if someone still lives in Iga or Koka but is hired without being noticed by others, then in this situation it is called In no Shinobi.

Upper tradition:
What follows is a deep secret. It may seem that only In no shinobi appears to be important, therefore, the question may be asked; for what reason should Yo no shinobi be used? The answer is as follows: if it is known to other provinces that you have hired a prominent shinobi, it will be difficult for other shinobi no mono to carry out any investigations in your province. For example,[11] it is just like when there is a good Meakashi[12] (police detective) in your area, who makes it difficult for thieves from other provinces to come to steal. If [shinobi] do try to investigate, they will show themselves and contact the prominent shinobi no mono of that area. This is just like the way Meakashi (police detectives) show themselves and if Meakashi from other provinces are following a target, they will come to get help from Meakashi of your area. Shinobi act in the same way.

In this way you can use the enemy spy who has contacted [the Yo no shinobi] as a Converted Spy 反間. Thus there are a lot of benefits in hiring shinobi no mono openly. Also, think on this, if you have covert shinobi as well [as open shinobi], and if you do not use shinobi no mono of Yo [for assignments] but send out only In no shinobi to carry out missions, then they will be able to do what they want and with ease.[13]

忍身持之事

Shinobi Mimochi no Koto

GUIDING PRINCIPLES OF BEHAVIOUR FOR SHINOBI

There are myriad oral traditions you should keep in mind that are different from the way of normal people in doing things or in the way of talking. Everything cannot be written down here and only important points are mentioned below.

What you should strictly restrain yourself from is telling a lie or being two faced. You should keep yourself from being dishonest or insincere so that others will trust you. You should make people think anything you say is true, so that if you talk falsely [when the time comes], others will think that what you say must be true.[14] You should keep as much contact as possible with other shinobi no mono who are positioned all over Japan.

You should make yourself known to as many people as possible,[15] as well as shinobi, and always take the trouble to send presents and greetings at New Year, midsummer and midwinter and keep in as close a relationship with people as possible.

Sometimes you will get military leave to visit your home and therefore you will travel at irregular times every year. Even if you do not have to take this leave, you should do. If you do it every year and everybody considers it is a usual thing, then when you need to go out for the purpose of investigation in emergency and discreetly, you can say you are taking leave to go home, which will have become a usual thing to do. This means that people will not realise your true intention and pay no special attention [to you leaving the area]. The length of time that you should be away may be up to fifty days most of the time, but the length can be varied, sometimes five to ten days, and other times it may be twenty to thirty days until you come back.

Supplementary to above:

While you are on leave to return to your province, you should travel around widely and make acquaintances in various provinces and also renew your friendships with your old friends, while doing this, you should investigate what the fortifications, roads, manners, customs and military forces are like in each area you are in. You should have a taste for some of the arts in order to make it easy for you to travel around. The most appropriate ones are poems, Renga verse linking and Haikai poetry 俳諧. The second most appropriate subjects are other literature and artistic interests [such as music, singing and dancing]. It is difficult for you to accomplish your purpose only with martial arts 武藝.[16] You should always be aware of this. Also, when you travel, you should go alone and come back alone without your actions being noticed.

You should be prepared to disguise yourself in various ways and according to needs, such as a Shinto priest, a yamabushi monk, an In-Yo diviner, a Komuso priest[17] etc. You should have a general understanding of what they do.

You should learn and have knowledge of medicine and pharmacy. There are other things to be orally transmitted.

忍衣類之事

Shinobi Irui no Koto

THE ART OF SHINOBI CLOTHING

When going to another province on a covert mission, you should sometimes intentionally show yourself as being from a different province, or sometimes pretend to be a local by mixing with local people. Also, sometimes you need to dress so that people cannot find out where you are from. You should do these things by choosing appropriate clothes for your purpose.

Generally speaking, you should wear normal clothes so as not to stand out and you should not be remembered easily by the clothes you wear.

Here is a tradition on how to change your clothing. For example, you venture out with a black kimono on top and kimono of fine patterns underneath, and when you return, you put on the black kimono underneath the one of fine patterns, which is now on top.

Remember, on a pitch-dark night you should wear something black and on a moonlit night you should avoid black.

忍心持之事

Shinobi Kokoromochi no Koto

HINTS ABOUT CLOTHING FOR A SHINOBI

You should learn well and have a full knowledge of how the local people of any given province wear clothes. This knowledge will allow you to look natural when you become mixed in with others and this will also help you to identify other shinobi from other provinces easily. There are more oral traditions on this point.

Remember, at the battle of Nagashino, Torii Sune'emon was undertaking his covert mission well but he wore the wrong kind of gaiters, which led to Kawahara Yataro detecting and capturing him.

忍可行秘法之事

Shinobi Ikubeki Hiho no Koto

SECRET MAGIC ARTS USED BY SHINOBI WHEN VENTURING OUT

Before you leave, you should pray intensely, see what divination reveals and make wishes for the success of your coming aim. If you have something that is a cause for concern, you should perform incantations, and pray and speak incantations enough [to dispel the problem].

Also, you should leave on a lucky day. If you have to go in an unlucky direction, you should take a measure for changing the direction.[18] Further details are to be orally transmitted.

忍出吉凶知事

Shiniobi Ideru Kikkyo wo Shiru Koto

THE ART KNOWING IF A SITUATION IS LUCKY OR UNLUCKY WHEN VENTURING OUT ON A SHINOBI MISSION

You should decide there is good luck or bad luck solely by noting and checking if you have a serene status of mind or if your mind is blurred and unfocused. You should not rely on any other way but this one.

香煙善悪之事

Koen Zen'aku no Koto

THE ART OF INCENSE SMOKE – IN CONNECTION WITH LUCK

If you cannot tell good or bad luck by the status of your mind alone, but you have a niggling cause for concern, you should perform divination in the following way:

In the first third of the Hour of Tiger, you should purify yourself, dress yourself properly, burn incense facing north and chant the following spell 180 times:

乾元享利貞[19]
Kan gen ko ri tei

Put your hands together, and make a bow to the gods of heaven and earth saying:

人生固陋ニシテ勝負吉凶ノ運ヲ知ラス仰キ願クハ天清雲ニ納受マシマシテ勝負吉凶ノ徴ヲ示シ玉へ
Jinsei Korou nishite Shoubu Kikkyo no Un wo Shirazu Aogi Negawakuba Ten Seiun ni Osame Uke Mashimashite Shoubu Kikkyo no Shirushi wo Shimeshi Tamae

> Human existence is stupid and obstinate and we do not know if the issue at hand will fail or succeed. Thus, I pray that the gods in the holy clouds will accept my wish and show me a sign to inform me if this matter will be successful or not.

Then, after this, burn incense, close your eyes and chant the spell *kan gen ko ri tei* 乾元享利貞 three times, then open your eyes and look at the smoke of the incense. If the smoke is rising straight up in one line, it is a sign of good luck. If the smoke is divided into two or three lines, trailing sideways, rising irregularly or unevenly or breaking up, then it is a sign of bad luck. More is to be orally transmitted.

忍道踏様之事

Shinobi Michi Fumiyo no Koto

THE SHINOBI ART OF WALKING

On a normal road, you should walk by stepping firmly with your heels first. In a house or construction, walking without sound is desirable. While on a muddy or slippery road, you should walk stepping firmly with your toes, so as not to easily slip. Oral traditions are to be transmitted.

One tradition says you should always try to walk on every road around various provinces as much as possible. If you cannot venture to far-off provinces, you should go and walk around at least your neighbouring provinces and have a thorough knowledge of their topography. This is called *how to tread a path* because you actually go and walk around the area. Oral traditions are to be transmitted.

忍路見之事

Shinobi[20] Michimi no Koto

THE SHINOBI WAY OF KNOWING ROADS

When you are going to where you have not been, you should have a guide draw a map for you to carry, so that you can go there by checking this map – this is done in case you cannot find anyone local to ask, more to be orally transmitted.

One tradition says that you should obtain the map or information for even far provinces by asking those who are from that place on the way [to the province]. This is called 'Michimi', which is 'to look over roads' and it will allow you to learn about where you cannot go by looking over these maps.

森林不審之事

Shinrin Fushin no Koto

THE ART OF IDENTIFYING SUSPICIOUS WOODS

When you go for a preliminary examination before your army march on, if there are woods where you think there may be an ambush, etc., then you should go in to the woods by stealth 忍, and one by one at intervals so that you can see each other.[21]

Everyone should carry a 'waist-musket'[22] and should shoot if he finds anything suspicious. If the first shinobi [in your group] is killed, the second or the third will spot this and report it by relaying it back one to another. In most cases it is shinobi who are sent to detect ambushes, that is why there is such a tradition as this one, more to be orally transmitted.

One tradition says, if you find suspicious woods, you should avoid them and pass by taking a longer way around. It also states that you should not get close to woods as ambushes must be waiting for you and that anyone inside will be taking aim with muskets. More details are to be orally transmitted.

昼夜川越之事

Chuya Kawagoe no Koto

THE ART OF CROSSING A RIVER IN DAYTIME OR AT NIGHT-TIME

To cross over a river in the daytime, you can; walk, swim, take a boat, walk over a bridge or cross on horseback; however, [during the day] you should not cross the river by using river crossing tools. When you use those tools, you should make a thorough investigation of the area and use them only after taking precautions. It is difficult to use the tools again. There is more to be orally transmitted.

At night time, you can use any tool and choose any way which allows you to cross a river with ease. Oral traditions should follow.

四足逢吉凶之事

Shisoku Au Kikkyo no Koto

THE ART CONCERNING CROSSING BY ANIMALS AND THE LUCK OR BAD LUCK IT BRINGS

If you come across Yo-animals coming from a Yo-direction, it is lucky; while In-animals come from an In-direction is unlucky. If the animal is crouching[23] while facing you, it is considered bad luck, so you should not venture that way. However, if the animal is lying down then it is good. Generally, if it is cattle, horse, deer, a fox or anything of that kind, it is good if it is male, but if it is a female it will bring bad luck. If an animal comes from your left side or from the south east, it is good luck while if from your right side or north west, then it is bad luck. Even if a mouse or weasel crosses the path in front of you, as long as it comes from the left side, it is good while if from the right side, it is bad. If you encounter something indicating bad luck, let others go ahead and follow them.[24] There are more oral traditions here.

諸鳥吉凶之事

Shocho Kikkyo no Koto

THE ART OF LUCK IN RELATION TO VARIOUS BIRDS

If a bird goes toward the enemy, it is a sign of good luck while if it comes from the direction of enemy, it is of bad luck. If a bird calls an odd number of times it is lucky, while if it calls an even number, then it is unlucky. More details to be orally transmitted.

諸虫心付之事

Shochu Kokorozuke no Koto

THE ART OF PAYING ATTENTION TO VARIOUS INSECTS

In older times, there was an excellent shinobi. One time he snuck up on a shinobi comrade's[25] house and when he went up to the veranda, he discovered [without seeing him] that the shinobi sleeping inside had just woken up and so he retreated. Next day, when they met, the ['sleeping'] shinobi said to the first one, 'Why did you not come inside last night as you came all the way to the house'. The infiltrating shinobi replied, 'I found that you had woken up, so I did not venture inside.' Then a third person hearing this asked, 'How did you two discover the presence of each other?' The shinobi who had been inside said, 'The insects in the garden gradually stopped chirping, first at a distance and then closer and closer, therefore I knew he had infiltrated.' The infiltrating shinobi said, 'Though there had been no sound previously, the sound of mosquitoes suddenly began, so I knew that he had woken up and come out of his mosquito net.'

This episode has been mentioned to remind you that you should pay attention to such things. More details to be orally transmitted.

死人逢吉凶之事

Shibito Au Kikkyo no Koto

THE ART OF LUCK IN RELATION TO PASSING BY A DEAD BODY

If the body lies with the face toward a Yo-direction then it is lucky while if an In-direction then it is unlucky – remember to be very careful. More details to be orally transmitted.

死人通様之事

Shibito Tooriyo no Koto

THE ART OF PASSING DEAD BODIES

[When you need to move past a dead body] you should turn it towards an In-direction and pass by it, however, do not move it toward a Yo-direction. It is a common rule that dead men should be laid face down and that dead women should be face up, it is unlucky if men are laid face up and women face down. More details to be transmitted orally. For drowned victims, this way is particularly important.

敵城用意知事

Tekijo Yoi Shiru Koto

THE ART OF KNOWING HOW THE ENEMY HAVE PREPARED THEIR CASTLE

You should investigate and always have information [on this subject] before an urgent situation arises. If the situation becomes urgent before you have acquired the information, you should [send someone to] investigate [their position] as soon as possible while preparing [for the upcoming conflict]. After the enemy become secure in their fortress, it is difficult to get information. More details to be orally transmitted.

敵逢陰所事

Teki ni Ai Kakuredokoro no Koto

THE ART OF KNOWING WHERE TO HIDE WHEN YOU ENCOUNTER THE ENEMY

It is too late if you try to find a place to hide after you have been detected by the enemy; tradition states that it is important for you to prepare a place you can hide beforehand. However, after you have exhausted every effort to hide yourself and have no way to hide any more, you should be resolute and prepare to die in battle instead of being captured. More details to be orally transmitted.

NOTES

1 Yin and Yang war curtains.
2 This same term is used to mean scouting.
3 This dramatic statement simply means that natural phenomena are beyond one's power to influence and that they may work in your favour or not.
4 This does not imply the shinobi is on horseback, it is a reference to how high the curtains may rise.
5 At times written as Innin and Yonin.
6 The additional information from the Yokan Denkai Kudensho scroll simply adds that the following are Koka skills.
7 The ideograms for lower, middle and upper – as translated here – change to 一傳 二傳 三傳.
8 Man of Koka, a name associated with shinobi.
9 Man of Iga, a name associated with shinobi.
10 This information is from Yokan Denkai Kudensho.
11 This example is from the additional information from the Yokan Denkai Kudensho scroll.
12 Detective or policeman.
13 Put in a convoluted way, this simply means hire open shinobi so that people know who your spies are – with the benefits described above – but also at the same time hire hidden shinobi and send them out on missions. Then people who are watching your forces will concentrate on the open shinobi, which will give the hidden shinobi freedom to work.
14 This is not a lesson in morality. He is saying that your cover should be a man who is known for not lying, however, this is done only so that people will believe that the agent always speaks the truth. This all used to support the larger lie that is the aim of the shinobi.

15 The point here is that an agent should be a socialite and have contacts with many people in many areas who are unaware he is shinobi. However, other shinobi will know the agent as a shinobi.

16 As a samurai the agent will be trained in martial arts, so here the text is saying do not just use the position of a warrior to make contacts, do not just interact with other people based on the study of weapons and war skills, using other arts to make acquaintances will widen the agent's circle of contacts.

17 A mendicant Zen priest of the Fuke sect, who wears a sedge hood and plays a shakuhachi.

18 To stay the night in a place situated at another point of the compass when your destination from home lies in an ill-starred direction.

19 This spell is found in the Chinese classic *The Book of Changes* (Yi-Jing) which is concerned with divination. It seems to mean the four virtues of Heaven, 'great and originating, penetrating, advantageous, correct and firm'.

20 The original ideogram here means 'same as above', therefore the title has been changed to 'shinobi'.

21 It is not clear if they mean the whole group of shinobi can see each other or if they are stretched out in a line where they can see the shinobi ahead and behind. The context would point to a line as opposed to a scattered group.

22 腰銃 - this term is unknown but implies a smaller weapon that can be carried on the waist.

23 The Japanese here means that the animal is lowering itself, probably in a threat posture against the shinobi; therefore, we have used 'crouching', even though not all animals can crouch.

24 It is unknown who the 'others' are. It could mean other travellers on the road.

25 同志ノ忍.

陰具巻

Ingu no Maki

THE VOLUME CONCERNING *IN* TOOLS[1]

六膳遣様之事

Rokuzen Tsukaiyo no Koto

THE ART USING ROKUZEN [SPIKES]

These should be made of forged iron, and measure 1 shaku and 2 or 3 sun long and be 9 bu in diameter, also they should be thinner and slightly flat at the end. They should be prepared in sets of twelve, which is why they are called Rokuzen.[2] They have a hole in them so you can carry them on a string. The primary use for this tool is for crossing over a wall or climbing stone walls. You should insert them one by one into a wall to make handholds at first and after climbing up [a short distance] you can use them as footholds. Continue inserting them one by one at higher points each time you move. Those following behind should climb up them as if they were climbing a ladder. More is to be orally transmitted.

Also if you need to stretch a rope out, you should stand [the spikes] at intervals and stretch rope across them. More details to be orally transmitted.

To cross over a roof, insert them between the shingles of the roof, more details are to be orally transmitted.

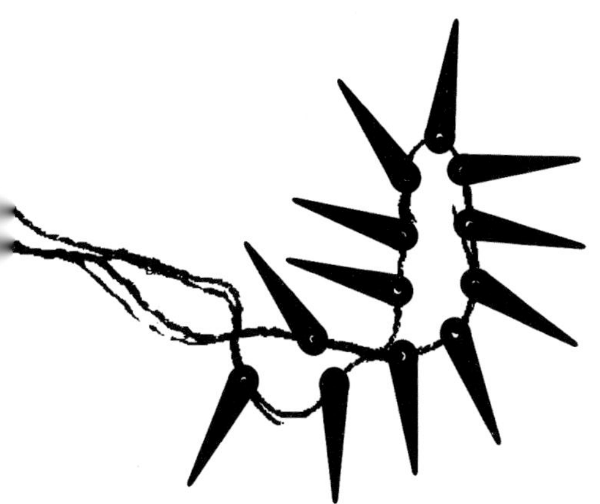

These tools can be used when you need to lock doors as well and again more details are to be orally transmitted.

Also, if they are made about 5 sun long, then they are called Metsubushi 'blinding weapons'.[3] It has been said that they were used to stab the enemy's eye when in close combat 組討. This shorter type can also be used for crossing a roof but are not used as footholds [for climbing walls]. If in urgent need, you can make them of high quality bamboo. This tool is versatile and you should be flexible on how you use it.

枢鎰遣様之事

Kururukagi Tsukaiyo no Koto

THE ART OF USING THE KURURUKAGI KEY

This tool should be made of forged iron and in the shape of a spindle. You can open most locks or latches with this. To open a lock, tear up paper and put it gently into the key hole and force the prongs of the shrimp [padlock with this paper], the prongs will be pushed together and become narrow, which will open the lock in most cases. You should ram [the paper] in with these Kururukagi shims, one at the top of the key hole and one at the bottom.[4] Do the same for handcuffs too. You should prepare four or five of these tools, including a thin and curved one and of various lengths, so you can use the most appropriate one according to the bolt or latch[5] you need to open. There are various ways to use this tool too. More details are to be orally transmitted.

微塵遣様之事

Mijin Tsukaiyo no Koto

THE ART OF USING DUST AND FINE GRAINS

You should take sand or ash and scatter it at a corner of a road or on a fork where people may not know which way to go to follow after you. This is done to inform those who come after you. More details are to be orally transmitted.

手楯拵様之事

Tedate Koshiraeyo no Koto

THE ART OF MAKING HANDHELD SHIELDS

This shield is to be used by holding it towards the direction from which arrows and bullets may come.

It should be a shallow box and the dimensions are:

- 1 shaku 3 sun wide
- 3 shaku long
- There is a handle on the inside of the back.

Cut well-tanned leather into pieces of 3 sun in width and 1 shaku in length. Place several of those pieces together overlapping with the strips [vertical. Start on the] right edge and sew them together with thin thongs of tanned leather. Put the now connected sections of leather into the shield box. 'Sandwich' this leather 'sheet' with thick whalebone on the outer side of the shield [the section that faces outward] and an iron sheet of 5 rin in thickness on the inner side in the box.

This shield will work regardless of what the distance is, [even against] muskets which have bullets of 3 momme 5 bu and that use as much as 1 momme 5 bu of gunpowder. More details are to be orally transmitted.

It is difficult to make the above shield in an emergency or in a short time. In such a case, soak a large and thick chopping board of willow wood and use it in the same way as the above shield. If you do not have a thick chopping board, fold up a thick wadded kimono and sandwich it between two boards [to make a shield]. More is to be orally transmitted.

菱結配様之事

Hishi Musubi Kubariyo no Koto

THE ART OF MAKING AND DISTRIBUTING CALTROPS

Caltrops should be made of iron and carried in a leather bag. If you think you are going to be followed on your retreat, you should scatter them and make your way out. More is to be orally transmitted.

When there is urgent need,[6] you should make nail shaped bamboo sticks that are sharpened at both ends and form them into a crisscross [and use them as caltrops]. More details to be orally transmitted. These bamboo nail-like slivers should be lightly roasted.

Alternatively, hammer three iron nails into a round board and have them fanning out and out of line.[7] These nails should have akamiso (red miso)[8] applied, roasted and smoothed before they are used. More details to be orally transmitted.

When you scatter them, you should be aware that if the enemy have difficulty crossing the area where you have deployed the caltrops, then you cannot pass by that way either. Therefore, you should only scatter caltrops in an area where you do not intend to pass through. More details to be orally transmitted.

沼浮沓之事

Numa Ukigutsu no Koto

THE ART OF MARSH FLOATING SHOES

When you pass through a marsh or deep paddy field, put together several pieces of bamboo in a grid-line frame [to put on your feet], use these so that you can cross over the marsh or field.

Alternatively, when you are the only one to cross over the marshy area, prepare [two] boards – each one approximately 3 shaku by 2 shaku – and secure a thong in the middle of each board. Then, put them on and tie them firmly [onto your feet]. This is done so that you can walk through the marsh or paddy. More details to be orally transmitted.

乱足沓之事

Midareashi Gutsu no Koto

THE ART OF CONFUSING YOUR TRACKS

There are three traditions for this technique:

Lower tradition:
If you want to avoid leaving [clear] footprints on the road, you should put your straw sandals on the wrong way round and walk for a while, then next put them back on the correct way and walk for a while longer. You should repeat this several times so that those who see the footprints later will be confused as to which way you were going, one way or the other.

Middle tradition:
When you sneak deep into house interiors, you should wear tabi socks that are made of leather and heavily soled and padded with cotton floss, this is so that you will make no sound. However, they should be made so that you can take them off quickly because if you are detected, you should take them off and fight.

Upper tradition:
When there is snow on the ground, you should step on your footprints in a crisscross manner[9] on your way back, so that the footprints cannot be identified. More details to be orally transmitted.

巻橋遣様之事

Makibashi Tsukaiyo no Koto

THE ART OF USING THE ROLL-AWAY LADDER

This ladder is used when you cross a stone wall or fence. Put a number of wooden rungs of about 5 sun in length onto two ropes made of cotton at intervals of about 2 shaku apart. First, have someone climb up to the top [of the wall] and hang this ladder from a fixed point, so that the rest can climb up by stepping on the ladder. Once you have all climbed up, you should retrieve the ladder by rolling it up. Bamboo can also be used for the rungs and if needs be, the rope can be made of the Ramie[10] plant. Thin sticks of forged iron can also be used as rungs. More details to be orally transmitted.

結橋之事

Yuibashi no Koto

THE ART OF THE TIED BRIDGE

This is used when you cross a moat. Prepare boards of 8 bu in thickness and 1 shaku 2 sun in length. Put holes at the ends of each board and join them together with cloth or cotton rope, up to the length you have estimated by observing the moat previously. You should have a good swimmer take this bridge and swim over to the other side and secure a spike[11] into the ground, then stretch the bridge across. The rest of your men can now cross over on this bridge. The rope to be used can also be made of the Ramie plant. Also bamboo can be used for the rungs; they should be 3 sun wide and 1 shaku long. The ropes should be laid in three lines, that is, both sides and the middle. More details to be orally transmitted.

早小屋之事

Hayagoya no Koto

THE ART OF THE FAST SHELTER

Taking your lodging at a local house is called Hayagoya. [This word] also means constructing a makeshift shelter with a bamboo or wooden frame and covering the frame with the rain capes that you have put together and then spending the night beneath it. Another method is to stand spears up, stretch rope around them, and cover the frame with Shibugami paper[12] or rain capes. The word Hayagoya also means to spend a night outside with just a reed hat and a rain cape, this is done to avoid dew and frost, even if it is a cloudless night.

道具持様之事

Dogu Mochiyo no Koto

THE ART OF CARRYING YOUR TOOLS

Make a bag of tanned leather and carry your tools in it. Make a separate bag for caltrops. More details to be orally transmitted.

There are a lot more tools than the above but they are described in another text.[13]

NOTES

1 Tools used for secret infiltration.
2 The ideograms used for Rokuzen 六膳 mean – 六 (six) 膳 (chopsticks or pairs). Today this latter ideogram is only used for chopsticks, however, traditionally it was used as a unit to count items that were thin and long and also for items that came in pairs. Therefore this is literally 'six pairs'.
3 Literally – 'something that blinds'.
4 The original says put the tool in from both sides. However, this needed expanding upon. The prongs of the lock splay outwards and catch on the housing, locking it in place. The paper, when rammed in, will force the prongs together and open the lock.
5 Kururu Kakigane クルゝカキカ子.
6 Meaning that there is not enough time to have iron ones forged.
7 The text states 'front, behind not in one line (or not straight)'. So this would be a small wooden disk with nails protruding out and not all vertical.
8 Miso is a form of soup in Japan, akamiso is a reddish soup. This appears to be red bean paste.
9 Most likely to step back in your own footprints when you leave.
10 *Boehmeria nivea*.
11 Most likely wooden.
12 Japanese paper strengthened with persimmon tannin juice.
13 Unknown writing, Chikamatsu was a prolific author and copies may have been preserved.

火巻

Hinomaki

VOLUME OF FIRE

筒火之事

Tsutsunohi no Koto

THE ART OF THE TSUTSUNOHI CYLINDER FIRE

There are numerous ways to make this, but it is difficult to make it when an urgent need arises because the ingredients are difficult to obtain quickly.

One secret method is as follows:

Prepare Nozarashi – which is old battered cloth thrown away and left outdoors. Make a rope of this cloth and wrap it with a straw bundle onto which you should apply a small amount of salt. This will remain lit for a long time. It is said this is the one that Tokimune[1] carried. The details are oral tradition.

Another method for a more refined version is: soak the above Nozarashi cloth in water. Put it on a stone and strike it with an iron hammer until it is frayed into small pieces. Dry the cloth sections in the sun and then grind them in a mortar until they are soft. Make a cylinder of silver or iron – the size should be the same size of the cylinder found in the game Sugoroku. Stuff the cloth flakes tightly into the cylinder with an iron hammer. You use it by igniting the cloth and by putting a cap on top. Wrap it with something so that it will not be too hot [when you hold it]. Put holes in the cylinder as needed. More details are oral tradition.

楊枝火之事

Yojihi no Koto

THE ART OF THE TOOTHPICK FIRE

This is also called the Kezuri Taimatsu – Shaved Torch.

The ingredients are as follows:

- Tinder from Kumano – 3 momme
- Saltpetre – 1 momme 5 bu
- Flint – 5 momme[2] jo-jo
- Japanese Royal Fern – 5 bu

- Sulphur – 1 momme 5 bu
- Camphor – 1 momme 5 bu
- Ash – 1 momme 5 bu
- Borneol – 1 momme 2 bu
- Saw dust of pine wood – 3 bu
- Pine resin – 3 bu

Powder the above, knead and solidify it with a decoction made from the bark of pine wood. If you strike [the solidified mass] with a knife, it will ignite. More is to be orally transmitted.

三重火之事

Mienohi no Koto

THE ART OF THE THREEFOLD FIRE

This is also called the Sodehi – Sleeve Fire. The ingredients are as follows:
Apply mercury onto three sheets of good quality Sugihara paper. Then crumple the sheets into a ball and light it. Wrap the paper ball with two or three layers of paper which has had alum applied to it. Carry this within your sleeve. More details to be orally transmitted.

籠火之事

Kagohi no Koto

THE ART OF THE BASKET FIRE

This is also called the Kago Taimatsu - Basket Torch. The ingredients are as follows:

- Camphor – 100 momme – this is the main combustible
- Borneol – 1 ryo – used to make the torch brighter
- Mouse droppings – 2 bu 2 rin – to help it burn slower
- Pine resin – 2 bu 5 rin – to sustain the fire
- Red clay[3] – 1 momme 5 bu - to make it burn bright but slow
- Cinnabar – 1 momme – to make it burn well – but only use this as coating

Powder the above and solidify the mixture with glue into a ball. It makes a good torch which will not go out be it rainy or windy. A ball of 100 momme in weight will burn well for a distance of 3 ri. Also, it illuminates a 3 cho square. Note, the fire will hold for four ri with a ball the size of 200 momme and will serve for 8000 people.[4] More is to be orally transmitted.

狼煙之事

Noroshi no Koto

THE ART OF THE SMOKE SIGNAL

- Wolf droppings – one third the amount of straw
- Pine leaves – one fourth the amount of straw
- Straw – a large amount

For example, for 9 momme of straw you should use 3 momme of wolf droppings and [roughly] 2 momme of pine leaves. This ratio remains the same for all amounts.

To make this smoke signal, dig a hole in the ground and put the above mixture into it. Then put a bamboo frame over it and light it. Next, insert a thick piece of bamboo in the middle of the fire which has had the bamboo joints removed; this is so that smoke will rise up through the bamboo. More is to be orally transmitted.

七里炬之事

Shichiri Taimatsu no Koto

THE SEVEN RI TORCH

This is also called the One Sun Three Ri Torch, or the Three Sun Seven Ri Torch. It is a torch which burns very slowly. The recipe is as follows:

- Saltpetre – 15 momme
- Sulphur – 15 momme
- Ash – 1 momme
- Horse droppings – 2 momme
- Camphor – 20 momme
- Mouse droppings – 2 momme
- Pine resin – 3 momme 5 bu
- Sawdust of pine wood – 3 momme
- Moxa – 1 momme 7 bu
- Hemp cloth – 1 momme 5 bu

打熢之事

Uchinoroshi no Koto

THE ART OF THE STRIKE FIRE SIGNAL

This is also called the Tenoroshi – Handy Fire Signal[5] – and is used for signalling at a short distance. Construct a long narrow box and create square, round or triangular holes within it, each of which has a glued paper cover. Arrange among yourselves the meaning of a message which is conveyed by square, round, or triangular. Then, if you remove the cover of the desired shape for the intended message you wish to send, your allies can see the light in that shape [and thus they can understand your message]. More details are oral tradition.

水松續之事

Mizutaimatsu no Koto

THE ART OF THE WATER-PROOF TORCH

- Saltpetre – 17 momme
- Sulphur – 12 momme
- Moxa – 3 momme
- Camphor – 12 momme
- Pine resin – 4 momme
- Ash of a hemp cloth – 7 momme
- Mouse droppings – 1 momme

Powder the above finely and ram it firmly into a bamboo cylinder. Then split the bamboo and take out [the dried mixture] and wrap it with paper so it is ready for use. More details are in the oral tradition.

切松明之事

Kiri Taimatsu no Koto

THE ART OF THE SLICED TORCH

This is also called Kiribi – Cut Fire.

- Iron filings – 15 momme
- Camphor – 2 ryo 8 momme
- Borneol – 5 bu
- Sulphur –1 momme 5 bu
- Saltpetre - 1 momme 5 bu
- Pine wood with extra pine resin – 1 ryo
- Ash of linen – 2 momme

Make the above into fine powder and add twelve momme of Kikumyoseki.[6] Solidify the mixture with glue and dry it in sunshine. If you strike it slowly with a knife, it will ignite. More details in the oral tradition.

手火矢之事

Tebiya no Koto

THE ART OF THE HAND THROWING FIRE

This is used to see inside enemy structures when it is a pitch dark night and to set fires. Cut bamboo which is as thick as you can manage to insert your little finger into and make sections of 3 or 4 sun in length. Attach good quality [black] powder used for torches at the end of this in the shape of an egg and wrap it with paper.[7] Next, cut bamboo or wood into lengths of around 1 shaku and sharpen [both] ends, and insert the longer stick into [the internal hollow of] the above grenade and launch it. More details are to be orally transmitted.

櫓落之事

Yagura Otoshi no Koto

THE ART OF TURRET COLLAPSING

- Saltpetre – 300 momme
- Sulphur – 52 momme
- Mercury – 240 momme
- Ash – 90 momme

How to use this is an oral tradition.

舟崩之事

Funakuzushi no Koto

THE ART OF SHIP DESTRUCTION

- Saltpetre – 50 momme
- Sulphur – 65 momme
- Mercury – 300 momme

When making this tool, put approximately 100 bullets which weigh 3 momme each into the mixture. More is to be orally transmitted

自見火之事

Jikenhi no Koto

THE ART OF DISCOVERING FIRE

You should not only rely on traditional ways of fire but try to discover your own recipes and test them. A little difference [in a recipe] can make a big difference [in the effects]. I decided to call this *The Art of Jikenhi* because it means to discover your own recipes and put them to the test. More is to be orally transmitted.

入死火之事

Irishibi no Koto

THE ART OF THE 'ENTERING DEATH FIRE'[8]

- Pepper
- Blue vitriol
- Bark of the red bayberry wood
- Blue centipede – to be charred
- Centipede – to be charred
- Newt – to be charred

Mix equal amounts of the above ingredients with gunpowder and make a ball of about 100 momme in weight. Make a hole for a fuse, light it and throw it into a structure or use it to capture those who are holed up in a position. There are important points you should be aware of on how to make and throw it.

人数崩之事

Ninju Kuzushi no Koto

THE ART OF DESTROYING AN ARMY[9]

- Saltpetre – 200 momme
- Sulphur – 45 momme
- Ash – 100 momme
- Japanese tiger beetle – 100 momme
- Pepper – to be charred – 100 momme

Attach the above mixture on to arrows to make fire arrows of them. With these fire arrows you can attack and drive an enemy army away, forcing them to withdraw.

NOTES

1 Hojo Tokimune 1251–1284.
2 The original text states 'jo-jo' 上々 which implies either 'more than' or 'excellent'.
3 土朱. This is ambiguous, most likely dry powdered clay.
4 This statement is unexplained and inexplicable.
5 The original simply says 'hand', however, it implies useful, portable etc., therefore we have translated it as 'handy' to combine the two.
6 Unknown, possibly a mineral of some kind, 菊明石.
7 Most likely the powder is wet or has melted resin mixed in. It is formed in the shape of an egg around the short bamboo rod and left to dry.
8 A form of tear gas.
9 A form of tear gas delivered via an arrow.

陰種巻

Kakureru Tane no Maki

HIDDEN ELEMENTS SCROLL[1]

The title should be read as Kakureru Tane[2] – materials for hiding. This volume lists the materials you need when on a covert mission.

貴妙散之事

Kimyosan no Koto

THE ART OF THE UNIVERSAL MEDICINE

When you need to infiltrate a castle or camp for a prolonged period of time, you may sometimes succumb to the heat or the cold, to sunstroke or to stomach-ache from worms and parasites, to headaches and many other such things. Therefore, you should carry good medicine for these diseases, medicine that works wonders.

Kimyosan means: a medicine 'for appreciating exquisiteness' and the word is made up of the following ideograms – 'ki' 貴 'to value' or 'to appreciate' and 'myo" 妙 'exquisiteness'.

'Ki' also has the same sound as the ideogram for 'wonders' 奇 and you should be aware of this.

One tradition says:
Kimyosan 奇妙散 is to be used when you are exhausted and if you take this medicine, you will be immediately invigorated. The recipe is as follows:

- White (Albino) snake [to be powdered?] – 1 momme
- Red rice, to be powdered – [?] momme
- Weak green tea – one cup

If you think the above is too strong, use this version:

- White (albino) snake [to be powdered?] – 10 momme
- Red rice – 100 momme
- Glutinous rice – 20 momme
- Japanese yam – 30 momme

Use the above.

井妙散之事

Seimyosan no Koto

THE ART OF BLINDING POWDER[3]

When you are detected by the enemy, you should scatter this around them or disperse it by fanning it, this is done so that the enemy will be bewildered and will allow you to escape. The ideogram for 'sei' [is used for its phonetic meaning] and is an alternative way to writethe ideogram 精 meaning 'minute' or 'essence'.

- Japanese tiger beetle
- Summer peach fuzz [shaved from the skin]
- Skin of the red bayberry [fruit]
- Pepper
- Hanabigusa[4]

Powder the above and keep it in a bamboo cylinder which is 4 sun in length. Put holes into the cylinder and scatter the powder over the enemy. You should keep the holes covered until you need to use it. This is also called *Yoshitsune's Mudra of Mist*.

How to bring someone back to their senses and to remove the effects is as follows: skin a mole and char the skin and make a powder of it.[5] Press the juice from apricots while they are unripe, and knead the charred mole with this juice and make a ball of the mixture. Have the person ingest it. This is called *Yoshitsune's Method of the Falcon*.

Another blinding powder tradition has a recipe called 'The Tradition of the Mist'.

- Pepper
- Ash of a cremated, rotten[6] dead body
- Thistle flowers
- Iron filings

Powder equal amounts of the above ingredients and carry it hidden within a bamboo cane or wrapped in cloth. If you scatter it over people, they will be blinded and confused.

一夢散之事

Ichimusan no Koto

THE ART OF DREAMING POWDER

This is a method to create a sleeping powder. Char the liver(s) of ray fish.[7] [Then when needed] put it into any fire or a lamp stand by twisting the charred material with your fingers, this is done so that people will fall asleep.

Another tradition states:

- Charred mandarin duck
- Incense powder

Mix equal amounts of the above and carry it wrapped in an old Tsuki [cloth]. Before you scatter it, you should insert pepper into your nose.[8]

Another tradition states:
Mix charred cats' tongue and mouse droppings and scatter them over the opponent [who is sleeping already], and in this way they will sleep deeper.

Another tradition states:
Dry Unagoji[9] in the shade and burn it [in the room of the enemy] and the people in that place will fall asleep.
Another tradition states:
Dry slugs in the shade and burn them [in the place where people are] and the people in that place will fall asleep.

未用散之事

Miyosan no Koto

THE ART OF USING POWDER TO KEEP SLEEP AT BAY

This is a powder that keeps people from feeling sleepy. Take the blood of a grey horse and mix it with Haraya [mercurous chloride] and carry it, and in this way you will not feel sleepy.

One tradition states:

- Asian ginseng
- Ziziphus jujuba
- White Tuckahoe[10]

Powder equal amounts of the above, and take it with a cup of weak green tea and you will not feel sleepy.

Another tradition states:

- Earthworms – to be dried in the shade – 1 momme
- Bagged[11] tea – 1 momme 5 bu

Powder the above and take it with hot water and kudzu starch gruel. Be sure to purify the earthworms with care. You should wash them so that they do not have any soil remaining before you dry them.

Another tradition states:
Char centipedes and mix them with glue. Make a ball of it and place it on the navel.

息合之事

Ikiai no Koto

THE ART OF CATCHING YOUR BREATH

This tradition is to be used when you are badly short of breath from running or working hard. You should make a paste of plum and sugar and carry it with you. Alternatively, grate unripe plum(s) and dry them on a board. Cut into small pieces and carry it with you. Or put a salted plum in a bag and put it into your mouth while still inside the bag.
Another tradition states:

- Loosestrife[12] – 10 momme
- Asian ginseng – 1 momme

Powder the above and keep it into your mouth.

Another tradition says:
Decoct the skin of pomegranates and soak paper in the decoction and dry the paper. Eat it when you are short of breath.

兵粮丸之事

Hyorogan no Koto

THE ART OF HUNGER PILLS

Normally you should carry food for two or three days. After you have infiltrated, you may not be able to return at once and may have to stay a while longer to investigate something. Therefore this tradition of hunger pills is used, as normal food does not keep very long.

- Black soybean – 5 go
- Sake – 2 go

Soak the beans in sake for one day and one night, remove the skin of the beans, dry them for half a day and carry them.

Another tradition states:

- Black soybean – 5 go
- Sesame – 3 go

Soak [the above] in water for one night and steam the above, three times. Then dry it completely and remove the skin of the two ingredients by hand and pound them. Make balls of the size of a fist and steam them from the Hour of Dog (about 8 pm) to the Hour of Rat (about 12 a.m.). Take them out [of the steamer] at the Hour of Tiger (4 am) and then dry them. If you eat one ball of the size of a fist, then you will not feel hungry[13] for seven days. Also, if you eat two it will last for forty-nine days, if three, for 300 days, if four, 2,400 days. Your colour and skin will not decline and your limbs will work as perfectly as normal. This above way is mentioned in the Nong Shu 農書, *The Book of Agriculture* [written by Wang Zheng in 1313].

Another tradition states hemp seed instead of sesame. Eat as much as you can and it will be the equivalent of eating twice or even three times.[14] Also, when you are thirsty, boil the above hemp seed in water and drink the water. When you have stomach problems, grind fresh *Malva verticillata*[15] with water and boil it. If you drink the water, your ailments will be relieved.

Another tradition states:

- Inner bark of pine wood – to be dried in sunshine – 1 kin
- Asiatic ginseng – 1 ryo
- Rice – 5 go

Powder the above three ingredients and make balls. Steam the balls in a basket steamer and harden them. This amount is enough for fifteen people not to feel hunger for two or three days. Also, it will improve your breathing.

Another tradition states:

When your rations are diminished and used up, you should gather spittle in your mouth, as much as possible and swallow it. Repeat this 360 times in a day and a night. This way you can survive tens of days. This method was tested by my master[16] when fasting for seven days. This is an amazing method which appeared in Juseihogen 壽世保元.[17]

The above recipes have deep secrets to be orally transmitted and you should discover them with your own resources and employ those you find effective only after testing them. As there are a number of similar recipes in the world, you should learn and test them and use those you find best, you should not just stick to the above.

夜討之事

Youchi no Koto

THE ART OF NIGHT RAIDS

There is no better way than night raids to achieve an easy victory over a large number with only a small one. By nature, you have to be a small number when infiltrating, therefore, night raids are most appropriate [for this reason]. However, the key to victory is attack by fire. It has been said since ancient times that the victory of the shinobi relies on their fire skills.[18] As well as muskets, you should be trained in every kind of fire skill[19] and to always create gunpowder and be thus equipped with it, so as to use it if an emergency should arise at any moment. This is why the traditions above in the volume of fire have been transmitted here. You should keep it fixed in your mind that victory for your clan will depend on night raids and attacks by fire;[20] therefore, you should continue with your training and improving your skills at all times. These [ideas of night and fire attacks] are the most essential points and they are to be orally transmitted.

NOTES

1. Literally, 'hidden – material – scroll'; however, it also implies tricks for hiding.
2. The alternative reading would be Inshu no maki.
3. Literally 'minute/energy – exquisite – powder'.
4. Most likely a type of grass that resembles a firework shape, literally 'firework-grass'.
5. Alternatively, skin a mole and char the body only.
6. It appears this is the cremated ash of a rotten, presumably abandoned, body.
7. *Batoidea*.
8. Possibly breathe in pepper through the nose or insert pepper corns.
9. Unknown, possibly, Unagigaji – *Lumpenus sagittal*.
10. 'White' is unknown.
11. An unknown, literally 'bag-tea', 袋茶
12. ミソハギ　クチヘキ之 Possibly purple loosestrife, *Lythrum salicaria*.
13. Probably meaning that your vitality will not diminish and that you will remain healthy.
14. The original is ambiguous.
15. A plant considered beneficial for constipation.
16. Unknown person.
17. A famous Chinese text on medicine written in 1615.
18. Hi no waza 火ノ業
19. Hi waza 火伎
20. This sentence can be read as 'the speciality of your clan should be night attacks and attacking by fire'.

極意天之巻

Gokui Ten no Maki

THE SECRET SCROLL OF HEAVEN

前中後之事

Zen Chu Go no Koto

THE ART OF BEFORE, MIDDLE AND AFTER

This is also called *Kako Genzai Mirai no Tsutae,*[1] *The Tradition of the Future, Present and the Past.* The future (Zen) means the time to come, for example; when you are planning to infiltrate a castle, you should evaluate the best way to infiltrate and give it deep consideration, this is known as Zen 前, the future, or before.[2] While you are infiltrating and performing your aim, this is the known as Chu 中 or the present. When you have completed your aim and return, it is known as Go 後, or after. Most people understand [how important] the principles of Zen and Chu are but there are few people who understand how important the idea of Go is. After, or Go, here means to make arrangements for the way for you to return [safely] and the measures needed in the event that you are detected.

1. Zen 前 is to prepare your measures, paying attention to everything.
2. Chu 中 is to perform the mission itself.
3. Go 後 is to think of the *after* and what comes following completion of the mission.

More details are to be orally transmitted.

神貴之事

Shinki no Koto

THE ART OF RESPECTING THE GODS

For what is beyond human power or wisdom, you must understand that you cannot achieve great feats without divine favour and protection from the gods. You can only accomplish marvellous exploits of tremendous wonder when you have divine protection. Remember that it is all thanks to divine power that you can escape serious danger, like when you tread on the tail of a tiger or when you narrowly escape the mouth of the venomous serpent. Therefore, shinobi no mono – of all people – should respect and value the gods to the fullest and have faith in them to the utmost.

More is to be orally transmitted.

Oral traditions:[3]
Of all people who are exposed to danger, nobody is in more danger than shinobi no mono. You should understand how dangerous the situation is by remembering the ideogram used for shinobi 忍 which is a heart 心 that is written under a blade 刃. Overall you should keep your mind faithful and never forget this above point even in peacetime. More is to be orally transmitted.

The ideogram for shinobi 忍 has a heart under a blade and this means that you must remain determined at all times. Everything you do should be done with a full resolve as if you are on the edge of a blade yourself. If you refrain from doing anything unrighteous or unfaithful but act with complete sincerity, you will gain divine protection. Anything you do should be based on a strong faith in the gods. An ancient poem says:

心ダニマコトノ道ニカナヒナバ祈ズトテモ神ヤマモラム
If your mind is singularly in accord with the way of Principles, the gods and the Buddha will protect you, even if you do not pray.

As in the above poem, if you keep your mind faithful and honest, you will be protected and aided by the gods. Some people say that the poem should continue with: *'it is even more so if you do pray to them'*. As shinobi no mono are those who face far more danger than anyone else, you should worship and respect the gods with concentration of mind, be honest and loyal to the bottom of your heart and always restrain yourself from falling onto an unrighteous or an unfaithful path. Generally, what a shinobi no mono says is esteemed even by lords [and those in command] and they may decide whether to advance or retreat depending upon what information is given [by the shinobi]. Thus, [what they say] is of extreme importance and no other people are as important as shinobi no mono. Therefore you should make it your principle to be honest and respectful, to worship the gods and Buddha and always try to conduct yourself properly.

[空]礫之大事[4]

Soratsubute no Daiji

THE PRINCIPLE OF THROWING STONES WITHOUT AIMING

The first ideogram in the title is wrong and the ideogram 空 meaning 'empty' or to 'feign' is correct, thus it should read *The Principle of Throwing Stones without Aiming.* When you infiltrate, you should know if a position is securely guarded or not by throwing stones as an experiment. If you throw one stone, those inside the position may make a fuss of it and then calm down. After they calm down, you should try one more time and see if they make a fuss[5] again or not. If they do not, it means they have let their guard down but if they do, it means they are still vigilant. However, equally, if they keep quiet when they should have made noise, then it is possible that they keep a strict guard and are well prepared and they may not make noise intentionally because they are aware that shinobi have come. In this case, they will not make any noise and let you in, then once you have infiltrated, they will attack. To judge if this is the case, you should always try to get information on what kind of people they are [beforehand].

This principle is not limited only to throwing stones but you should apply this to everything. For example, you should set fire around the target early in the evening and see how they will react or if they make a fuss about it. Or you should set up a lot of gun shots and war cries at a distant place to see how they will react to this. Alternatively, you should put torches in the mountains so as to look as if you are coming in a larger number. You should make up unusual or strange events in various ways to see what their disposition is. All these concepts come under the name *The Principle of Soratsubute* ソラツフテノ大事. More is to be orally transmitted.

水中之大事

Suichu no Daiji

THE PRINCIPLE OF CROSSING WATER

Although there are various skills, [the best way is the following]: you should use a good swimmer from all of your men and have him swim across to the other side with rope and secure the rope on the other bank. Then you should send strong men across the water and have them strengthen the securing point so that it will not give way and have more ropes stretched out, this so that you will be able to cross the water via rope. The above method is a certain way to cross a river, so take note that other various devices or gadgets are sometimes risky. More is to be orally transmitted.

Another tradition states:
When you wade through water in the middle of winter, it is a common thing to have a sip of the river water before you go into the water, by doing this you will not freeze.

夢枕之大事

Yumemakura no Daiji

THE PRINCIPLE OF THE DREAM PILLOW[6]

There are various ways to stay [vigilant] for certain lengths of time, but you cannot rely on them completely. So you should take turns to sleep in shifts. While you are asleep, if someone keeps awake and on watch know that you are securely guarded. This is also called; *Nete Nezare no Tsutae* 子テ子サレノ傳 *The Tradition of Not Sleeping While Sleeping*.

大軍中通事

Taigun no Naka wo Toru Koto

THE ART OF INFILTRATING A VAST ARMY

The trick of this art is to [infiltrate] a massive army. As the intended army is massive, it is actually easier for you to become mixed in with the force. If the force is a small number, they know each other very well and you will be detected easily. However, in a large army, they are usually a composite of troops from various provinces, so no matter how strictly they try to put military orders into effect, it is difficult to have such orders carried out thoroughly without exception [and gaps will appear]. This is the reason why you can easily get mixed in and move around. You may think it is difficult to move around in a large number of people but this is not the case. The more people there are, the easier you can move around. You should know thoroughly and copy the manners and customs and any other points concerning the province [that the army came from], so that if questioned, you will be able to give an evasive answer.

This principle can be used not only for going through a massive army but any other situation where you consider it to be of use. If there is someone seeking your life in revenge, it is easier to hide yourself in a big city like Edo or Osaka, or other big castle towns. It is difficult for you to hide in a small place. More is to be orally transmitted.

寝屋之大事

Neya no Daiji

THE PRINCIPLE OF SLEEPING QUARTERS

There are three traditions which are as follows:

1 Those who are vigilant about making everything secure and close all the openings of their bedrooms so that the enemy shinobi cannot infiltrate. More details are to be orally transmitted.

2. For the lord's bedroom, be sure to close off any space under the floor. In wartime, you should have an all-night vigil under the floor.[7] Alternatively, the floor should be made so low that nobody can crawl in to the space. Even if this is done, you still should have the lord's bedroom in the middle of the mansion and keep night vigils on all four sides of his room. More details are to be orally transmitted.
3. Once you have infiltrated close to the enemy lord's sleeping quarters – enough so that you can break into the room – you should break through the doors or paper-covered sliding doors and make a sudden surge into the room, without worrying about making noise. [Making noise at this point is not an issue] as all that matters is reaching this point, as it is not easy to get this far. Now that you have got that far, the only thing that matters is speed and it is essential to attack quickly. No matter how vigilant the guards are inside his quarters, in all probability they will only number two or three but not many more. Whereas you are usually between seven and ten people when you make an attack on a bedroom. Therefore, it is essential for you to be determined to die and not to concentrate on surviving. If you succeed in killing the enemy lord, it is a great achievement even if you are obliterated in the process. All you need to do is overwhelm them at once and as quickly as lightning. Also, you should be prepared to kill him by shooting quickly with a musket.[8] More details are to be orally transmitted.

忍顕大事

Shinobi Araware no Daiji

THE PRINCIPLE OF SHINOBI IDENTIFICATION

You should not carry letters which will expose anything if they are read after you die. You should take measures before you infiltrate any position so that no one will know who you are and where you are from when you are dead, or even for which lord you work for as shinobi. The moral of this tradition is that you should be determined to die if your secret is detected. If you try to survive at all costs after completing your mission, it is risky, therefore, you should be resolute.[9]

身陰之大事

Mi wo In no Daiji

THE PRINCIPLE [OF RETAINING] AN IN-BODY

Your body should be determined to die and your mind should be fully aware.[10] You should make your mind active when you infiltrate because if you become mentally trapped, your mind will be dead, and your skills will be dull. When your skills are dull, you will be detected. Remember; your body and mind are two different things. You should make your body In 陰, that is, determined to die, but keep your mind Yo 陽, that is, you should stay resourceful and strong-willed and aware.

木之葉隠之大事

Konoha Gakure no Daiji

THE PRINCIPLE OF LEAF HIDING

The leaves of one thousand different trees are not all the same but are all different in one thousand ways. Just as they are different, you should be extremely flexible on how you infiltrate, which will be determined by the situation of the position you wish to infiltrate. Also, disguise yourself as someone who is expected to be let in.[11]

Since time immemorial, pine trees have had pine leaves and willow trees have had willow leaves, thus it is essential to behave as the enemy behaves. More details to be orally transmitted.

NOTES

1. 過去現在未来.
2. The time before a mission when you consider that which is to come.
3. Additional information from the Yokan Denkai Kudensho scroll.
4. The original ideogram from the manual does not appear in any dictionaries and the text itself states that the ideogram used here instead is the correct one.
5. If they come out to investigate.
6. 'Dream Pillow' means to appear in a dream.
7. A defender lying in wait under the house to stop any intruder.
8. 'waist-gun'.
9. The aim here is not to die, however, if a shinobi accepts death he will approach his escape in a different manner.
10. There is a small transcription error here that we have corrected.
11. It is better to be someone who is expected to have access to a position than be a person who needs to creep in and, therefore, you should have a disguise which allows you access.

PART II

AN IGA AND KOKA COMMENTARY ON SUN TZU'S 'USE OF SPIES'

PREFACE TO THE YOKAN DENKAI MANUAL – A STUDY ON THE USE OF SPIES

Master Yasutaka [who is Kimura of Koka] said that in principle the utmost and deepest secrets in the traditions of shinobi are all contained in the chapter 'The Use of Spies' in *The Art of War* by Sun Tzu. If you do not study this text in every detail and thoroughly and if you think what is essential for a shinobi is a diverse range of 'wondrous' skills, such as crossing a moat or climbing a wall, then you will be totally mistaken and you will be putting the cart before the horse.[1] As written down in this scroll, [Master Yasutaka of Koka] gave lectures on the true and deepest principles of the shinobi referring to the five types of spies [as written by Sun Tzu]. This is because he thought that you do not need to learn any other materials if you fully understand the text of Sun Tzu and master the truest principles and reasonings of the shinobi. I [Chikamatsu] wrote his teachings down here with respect.

Years later, I had the chance to meet Master Yorihide [of Iga] and commenced under his instruction. The master was teaching on the Iga traditions of the shinobi for certain reasons.[2] He said that you should regard the chapter 'The Use of Spies' by Sun Tzu as the fundamental and deepest essence of the shinobi, and he often lectured about the above text, a lecture I recorded.

Now that I have obtained and read the manual *Son Shi Kanrei* 孫子管蠡 written by Master Saigyoku [of Naganuma Ryu] I have found this text tallies perfectly with the traditions of Koka and of Iga. Therefore, I have compounded this text with the secret traditions from the two masters [of Iga and Koka] and I recorded it all. In doing this, I only recorded here the differences between the two but omitted those points which completely match with the text. Those who are to study this text in future should be aware of that.

Written in 1736; in the year of Hinoe and of the Dragon, in the twelfth month and on a day of the moon when in its last quarter.[3]
Written by Chikamatsu Hikonoshin Fujiwara Shigenori
In the south side of the Renpeido war school, to the south of the castle of Owari 尾府城

This manual consists of sections from the following texts:

The *Sonshi Kanrei* 孫子管蠡 – A Study of Sun Tzu's *The Art of War*[4]
From Bushu province
Collected and annotated by Saigyoku-ken Saeda Masanoshin Toyohara Nobushige Sensei [of Naganuma Ryu]

Also:
Supplements from *Koga Traditions* as lectured by
Kimura Okunosuke Fujiwara Yasutaka Sensei in Goshu.

Supplements from *Iga Traditions* as lectured by
Takenoshita Heigaku Minamoto no Yorihide Sensei of Mikawa.

Compiled and recorded by
Chikamatsu Hikonoshin Fujiwara Shigenori of Bishu.

THE PREFACE TO THE YOKAN RIGEN MANUAL – TRADITIONAL SAYINGS ON THE USE OF SPIES

Although the Yokan Denkai manual was written [damaged text], this further book has been compiled so that younger people[5] who cannot read [Chinese text] can gain an easier understanding and also it has been written to aid new learners of this art.

Written in An'ei 4[6] (1775) the year of Kinoto and the Ram on the fourth day of the second lunar month.

In our country the art of the shinobi was created by deities in ancient times and not transferred from foreign lands.

Concerning the origin of shinobi, there are two theories, that of Koka and that of Iga, explanations of these origin stories are omitted here as they are described in the Yokan Denkai manual.[7]

The writing of Sun Tzu was transmitted from China and the traditions of the five types of spy correspond with the traditions used by shinobi in our country. Therefore, those who learn the art of warfare have studied this text in depth. Those shinobi no mono from the two lines of Koka and Iga studied this chapter on the use of spies very closely and examined it deeply, and thought of this [text] as the most fundamental and essential of things – they studied this until the Middle Ages 中比. However, these days, the art of the shinobi has been reduced to an art which only deals with: crossing over moats or rivers, climbing up steep places, crossing over marshes, crossing over a high wall, changing appearances and deceiving common people. It is put in the same category with so called ways of Maho-magic 魔法 or Yojutsu-black arts 妖術 and has become humble and is to be performed by common people 凡下ノ者. As a result of this the speech, behaviour and performance [of today's agents] is poor and unskilled and very few of them know the deepest truths of the way of the shinobi and nor have they mastered it. Consequently, if performers of this art are asked [to perform their skills], there are few who have mastered it and are good enough to fulfil their role and complete their aims. Therefore, I am writing this Yokan Rigen manual using common and colloquial language to provide a guideline for the reader.

THE USE OF SPIES – THE THIRTEENTH CHAPTER OF *THE ART OF WAR*

The Sonshi Kanrei manual says:
Spies 間 are called different names according to the place.
In our country, they are called shinobi 忍.

The Yokan Denkai says –
Iga traditions say:
They are called shinobi because they make the activities they do mysterious and secret.

Koka traditions say:
They are called shinobi since they hide themselves from the ears and eyes of people.

The deeper traditions from Koka say:
According to one secret theory, shinobi is from the ideogram in the word 'kan'nin' 堪忍 meaning 'patience'. They are called shinobi because they endure the unendurable, and any difficulties or hardships that are imaginable.

In our country, the origin of shinobi is recorded in the upper scroll of *Records of Ancient Matters*, here the god Susanoo changed Princess Kushinada into a magic comb and put the comb in his hair, this comb was called Yutsutsumagushi. This is when [shinobi] first came into use – this is an Iga tradition.

The god Takamimusubi sent a bird called Nanashi no Kigishi which means *a common pheasant of no name* to investigate the situation [of how the god he sent years before was behaving on earth]. This is the origin of the shinobi – this is a Koka tradition.

I [Shigenori] think the tradition most appropriate is the one which says its origin is the myth in which Prince Susanoo transformed [the princess]. This is because in the other tradition, the bird Nanashi no Kigishi is the *nameless pheasant*, shot and killed with an arrow, so this theory is not appropriate. It is also inappropriate to say that someone from low birth[8] who has no name was the ancestor [of the shinobi]. Also, the bird was killed by an arrow before it fulfilled its mission so this [version] should be avoided. The other myth where Prince Susanoo changed himself into a woman[9] and killed Yamata no Orochi, a giant eight-headed snake, is when all the arts of war and divine tactics were originated and have been used for all generations since then.

When Emperor Jinmu went on a military expedition to the east, Shiinetsuhiko[10] disguised himself as an old man, and Otoukeshi[11] changed his appearance into an old woman and passed through a massive army and successfully brought back clay from Mt. Amanokaguyama. Also, the Emperor Keiko[12] disguised himself as a woman and defeated Yasotakeru.[13] These events all have their roots in the divine plans of the god Susanoo no Mikoto [who transformed and disguised the Princess]. Thus, the amazing tactics of the shinobi should be respected and valued. However, the task [of the shinobi] is now regarded as a humble occupation and it is thought to only be done by those who are middle or lower rank samurai; this is wrong. If someone says that shinobi no jutsu is not useful enough, then they know nothing of the art of war[14] and there is no need to comment about their opinions.

According to the Oboshi Kugi 大星口義 text written by Master Saigyoku, there is a place upriver and along Niu River, where they worship the gods in heaven and deities on earth [and talk of the connection between gods and humans].[15] This may seem irrelevant and that it does not pertain to this subject, but it is mentioned because spies should be aware of this. The principle of *A Reverence for Gods* 神貴ノ大事 from Koka traditions is very similar to this and you should have respect and faith in these things.

In modern times, [shinobi] are called suppa スツハ or yato-gumi,[16] and the ideogram 竊盜 is used for the word shinobi.[17] These [secret skills using this ideogram] are often secret traditions taught by various military schools. However, the true traditions of shinobi are not only the skills of thieves and those which involve vicious plans and cruelty, do not be misled by this attitude.

Master Saigyoku [of Naganuma Ryu] says that the ideogram Kan 間 – 'between' – is used because spies move *between* your enemy and your allies.

Koka traditions say:
Shinobi are the ears and eyes of the general. They know all things by looking and listening, thus, they serve as the ears and the eyes [of their leaders]. If they make mistakes in those things which they see or hear, then an entire army may be defeated and the state may be ruined. Therefore, this job influences the entire war or campaign, that is why we have the secret tradition of Mitsume and Kikitsume.[18]

Iga traditions say:
If you conduct a war without using shinobi, you will not be able to win even one battle out of a hundred. Shinobi are the ears and the eyes of the general. If you fight without the ears or the eyes, it will be a so-called *blind battle*. Thus, the very key to victory is shinobi. Be it in Japan or in China, excellent generals who used shinobi achieved great feats, while those who did not were always defeated. Therefore, as a result there is no general who does not use shinobi.

Koka traditions say:
Be it in Japan or in China, there have been a countless number of people who talk about warfare, but only Sun Tzu talked about spies. Other people [when they do talk of spies] only regard them as using vicious skills or as thieves or with the principle of using violence against violence, therefore it is not worth discussing their points of view. Although there are hundreds of military schools in our country, none of them knows fully of the use of spies and none [of these scholar-tacticians] discuss this matter in detail. This is because they study Sun Tzu's chapter 'The Use of Spies' without care and are not correctly acquainted with this matter. Therefore, you should study Sun Tzu's chapter as much as possible, as the principles essential for the victory of the entire army lie in this section.

The Yokan Rigen says:
Traditions on the Meaning of the Name Shinobi
In China, [shinobi] are called Kan 間, Cho 諜, Saisaku 細作, Yutei 游偵, Kojin 行人, Soshi 操詞 or Shoryaku no Shi 抄掠ノ士 'robber-knights'. In our country [Japan] they are called shinobi 忍.

Iga traditions say:
They are called shinobi because they are secret and hide secrets in mysterious ways.

Koka traditions say:
They are called shinobi because they hide themselves from the eyes and the ears of other people.

Deeper traditions say:
Shinobi is from the word 堪忍 perseverance. Shinobi no mono cannot achieve great feats unless they endure hardships which are unendurable, fulfilling the most difficult and painful of tasks. This job cannot be fulfilled if you are not determined or if you are impatient. Thus the ideogram 忍 is read as shinobi.

神貴ノ大事

Shinki no Daiji

THE PRINCIPLE OF REVERENCE FOR THE GODS

Koka traditions say:
Shinobi [skills] of Japan were originally performed by gods and did not originate from human concepts. These days [the task of the shinobi] is regarded as a humble job to be done by the middle and lower level samurai, but this way is wrong. Some people say that shinobi no jutsu is not good enough to be used practically, however they do not know much of the art of war,[19] thus it is not worth discussing.

The master says:
[The way of the shinobi] was ordained by the path of gods but cannot be achieved without training yourself in human-based skills. If you only rely on Shinto, 'the path of gods' and lack human skills or have no concern with human matters, then know that you do not respect the way of Shinto correctly. However, if you try to perform human skills with a selfish motive, then you will not have divine protection. Therefore, a spy should be very careful about how they approach the application or disregard the mix of the two. Basically, all which is performed and achieved is a human matter but these human skills should be performed with the utmost of faith. These affairs and the path of gods are not two separate issues. For example, the 'original chi' lying deep inside humans is considered to be connected to the path of gods while, medicine and self-cultivation 修養 are considered human matters. You should consider the above deeply and be very careful not to be thoughtless so that you can achieve great feats; more details are in oral traditions.

These days [shinobi] are called suppa or Yato-gumi. Also the ideograms 竊盜 are used for it and pronounced as 'shinobi' and are often found in the secret traditions of various military schools, but the skills [described in these manuals] are mostly vicious plans or cruel skills of evil men and not the true traditions of shinobi. Therefore, you should not be confused about this issue.

THE USE OF SPIES

Those who go between the enemy and their allies and investigate the enemy status, hiding themselves are called Kan 間 – which means between. Yo 用 [from the title word Yokan] means 'to use' and therefore this is what Yokan 用間 means – 'the use of spies'.

There are [two kinds of] spies, those of importance and those of lesser importance.[20] They sometimes disturb enemy politics and twist the enemy so that their military plans are confused and wrong; also they move around spreading rumours and perform assassinations. Excellent generals use them well to realise victory. When generals move forth with their armies, they need to know precisely about the following matters; what the adversary's political situation is, their military rules, what their system of punishment and reward is like, their code of orders, amusements and arts, tools, the provisions they have in reserve, army and troop numbers, the strong and weak points in everything, people's disposition, their manners and customs, if they are brave or cowardly, wise or ignorant, topography, if the four classes of people are rich or poor, etc., and the ease in which they can dispatch their army. All these kinds of information should be obtained by way of spies.

These types of information should be investigated by using spies before any situation becomes critical. Also, those leaders who are extremely vigilant always conduct a close investigation of every daimyo, even those in distant provinces, however, if they cannot be reached then these generals should fully investigate at least those clans around them. They usually try to remain informed about the routes in which they could enter [the enemy province], the topography such as the mountains, how deep or shallow the rivers or the seas are, etc. This is done so that there will be no problems if an emergency arises. It is difficult to obtain such information without the use of spies.

Also, although it is not correct to execute vicious plans or evil stratagems, sometimes it is unavoidable in tactics, as you want to have the smallest number of casualties and gain victory without difficulty. Also, if you stop using spies, it will provide the impetus for your own downfall via enemy agents. Remember, you will not able to win without the use of spies.

Koka traditions say:
Shinobi are the ears and eyes of the lord and the role of the ear and eye is to hear and observe and recognise things. If they see or hear something incorrectly, it may cause the defeat of the army and the ruin of the country. Therefore, the essence of warfare lies in the task [of the spy].

Iga traditions say:
If you fight a war without using shinobi, you will not able to succeed in even a single conflict out of 100 battles. [Shinobi] serve as the ears and eyes of the commander in chief. If you fight without ears and remain deaf, then this will be a so called 'war of the blind.' Therefore, shinobi are the key to the victory of any army. Excellent generals, be it in Japan or China, achieve great exploits whenever they use them. When they do not, they are defeated without exception. You should think closely of the victory or defeat of those battles of old.

Master [Saigyoku of Naganuma Ryu] gives a grave warning about the use of shinobi and that is, be warned that there are a myriad people who utilise shinobi and fall into vicious plans, evil and cruelty. Therefore, there are two ways of using them, *thoughtlessly* and *thoughtfully*. The *thoughtless* way is if you use them in a careless manner, there are a lot of old examples where [the incorrect use of spies] overturned a war and ruined a country. Opposed to this, using them *thoughtfully* means to use them for grave matters and with special care in all details. A general should take the righteous and faithful path as a basis, think of only the 'true mind' and of reality [not that which is only on the surface] as essential and do not rush for fame nor be blinded by greed, also, choose well the path where a spy should be used, do not have an egotistical mind, have the justice of heaven contained deep within the heart, keep self-restraint, process deep speculation, stand in awe [of the gods], make no mistakes; [if this is achieved] then the use of shinobi will lead to a completely righteous path. This is why Sun Tzu teaches about this matter, so that it will help you with strategy and the ability to attack or defend and realise victory without difficulty.

Koka traditions say:
There are a numerous people who discuss warfare, be it in Japan or in China. However, Sun Tzu is the very first of the great commentators who talked about the use of spies. Other people sometimes talk about spies but all just consider them to have vicious skills and that they are evil men and use a violence-for-violence principle, this it is not worth discussing.

Although there are hundreds of military schools in our country, none of them know much about the traditions on the use of spies and none can discuss this topic in detail. This is because they do not study Sun Tzu's writing on 'The Use of Spies' in depth and are not thoroughly familiar with the text. Hence, you should try to study it very closely.

THE TEACHINGS OF SUN TZU'S THIRTEENTH CHAPTER

Sun Tzu says:

> Raising a host of a hundred thousand men and marching them great distances entails heavy loss for the people and a drain on the resources of the State. The daily expenditure will amount to a thousand coins. There will be commotion at home and abroad, and men will drop down exhausted on the highways. As many as seven hundred thousand families will be impeded in their labour.

The Yokan Denkai says –
Iga traditions say:
If spies are used, it will allow you to obtain an easy victory, while if not, you may lose in a battle which you could win. This is why Sun Tzu suggests you should use spies. When you make a suggestion to a lord or general that they should use spies, in order to convince them, you should first talk about the aforementioned high expenditure as discussed in the Use of Spies [by Sun Tzu]. Your argument will take them by surprise, next you should immediately continue to talk on the diversity of the use of spies and those [positive] results which they can yield, do this to convince them on the subject.

The Yokan Rigen says:
When you take a hundred thousand men to attack a distant province [in China], know that it will cost approximately one thousand gold coins per day. This is the same as is estimated in our country [of Japan] today.

The estimated amount of rice required for one hundred thousand men a day is one thousand koku, when this is divided by the number of the people, it equates to 1 kin for each man, this means giving them 5 go for the morning and evening. Also the estimated amount of miso paste required for such a force will be 40 koku, this is based on of giving ten people 4 go twice a day, that is in the morning and in the evening. The number of horses will be around one thousand and therefore 20 koku of soy beans are required each day, with the estimation of 2 sho to be given to each horse every day.

Marching a Thousand Leagues

The foreign country [of China] is so broad that [Sun Tzu's text] states it as 'one thousand ri'. However in Japan, there is no such land space which spans a distance of one thousand ri. Nevertheless, you should remember Lord Hideyoshi's expeditions went to the provinces in the west and to the east, continuing on afterwards to Korea, and as far away as the provinces of Oushu and Aizu. Therefore, it is difficult to estimate how far such a distance is, thus [the text] mentions one thousand ri as an example.[21]

The Cost to Farmers

In China, the soldiers were farmers and they had to stop farming when they went to war. That is why the title says 'the cost to farmers'. In Japan, farmers are not used as soldiers but as labourers so they work as labourers and therefore [in war] they cannot farm their own rice fields or other's fields, thus war is also [an indirect] cost to farmers.

A Drain on the Resources of the State

The above paragraph talks about the cost to farmers, that is, the common people. This passage talks about the huge sums that common people offer [as war tax] to the high and the noble. It states the use of one thousand gold coins each day, which is the same as mentioned above.

Commotion at Home and Abroad

This means there is not only commotion with men who are outside but also with women who stay at home. Even today in peacetime, when people from a province leave Edo,[22] their wives, mistresses, or even maids are disturbed and cannot do any of the things they usually do. It is even more so if the distraction is preparation for war.

Men Will Drop Down Exhausted on the Highways

On a march, troops and horses are exhausted so the people of the houses and lodgings along the way have to take care of them and they cannot concentrate on their main tasks and will therefore enter into financial difficulty.

As Many as Seven Hundred Thousand Families Will be Impeded in their Labour

In China, the fields are divided according to the system called *seiden* 井田. In this system fields are divided just like the sections of the ideogram 井. This means that fields were divided into nine sections [as in the ideogram] and only the centre section was offered to the governor [farmed for tax]. The other eight sections were taken by eight farmers. Once a conscription had been issued, one person out of the eight becomes a soldier. The cost was paid by the other seven people so that the soldier in question could prepare [for war]. As a result, not only the one [soldier] but also the other seven have difficulty, as it is hard for them to farm efficiently. Therefore, if an army of a hundred thousand people is assembled, then seven hundred thousand families will have difficulties.

Today in Japan, bushi warriors and commoners are clearly differentiated and if a war takes place, commoners do not become soldiers and it may seem as if the above situation [in China] will not come about; but the cost will actually be borne by commoners. However, although we do not have the same allotment system as the *seiden* 井田 system of China, it is all the same in that commoners have to pay gold or silver for war. Daimyo of the present time, if they do not have enough money, assess peasants and merchants [for tax] to cover any shortages. Especially in the turbulent period of the past, where [Daimyo] had to launch armies frequently, there was no other way to fund this than collecting money from peasants and merchants, meaning that the people in a given province were in dire straits and serious uprisings and difficulties took place, as was the case with the collapse of the Takeda clan. Those who would be well prepared, be it of high or low rank, save money for war funds, however, such savings will be enough for the initial costs of war but soon the expenditure will outrun the savings and all of the costs cannot be met, especially if they have to dispatch an army each year. Also, peasants will be requisitioned as labourers for construction and will have no time to farm, merchants will have difficulties as people do not spend money on pleasures or extravagant amusements, and their business will dry up. For craftsmen, as the demand for luxury goods or artistic and exquisite work abates, they have nothing to do and business fails. [Therefore in the time of war] all the peasants and townspeople were reduced to poverty. On top of this, they were taxed, in gold, silver or rice, for extra military expenses, in addition to the usual taxes and this leaves people in significant hardship. Generals and officials should be extremely thoughtful and concerned about this point and should not think it easy to raise an army.

I am discussing this issue in depth because you can take victory without exception if you know the enemy status, be it *substantial* or *insubstantial*, strong or weak, etc., but remember it is difficult to view an internal situation if it does not appear on the surface. If you get information on what they are truly hiding, you can take appropriate measures and win a thousand battles with ease. It is only through the use of spies that you can know the confidential matters or secret plans of the enemy. Thus, if you obtain the information by the use of spies, you will be able to make the enemy surrender without smearing swords with blood or if you do have to fight, you will win with ease and with the least number of casualties because you can choose the easy path to victory. Taking this into account will enable you to reduce the cost for both public and private sources, spare the worry of death, release people, profit the strong, enrich the province and all these factors lead back to the quotation [in Sun Tzu's text] 'As many as seven hundred thousand families will be impeded in their labour.' All these things are enabled by the use of spies. This has been written to encourage the use of spies and it has been explained in detail to show just how much loss an army would suffer [without them].

Iga traditions say:
Generally, if you use spies, you can win with ease. If not, you may lose a battle that by rights you should have won. This is why Sun Tzu recommends the use of spies. To convince lords and generals of the benefit of their use, he mentions an immense amount

of expense so that it will surprise leaders. Then he suggests that a small amount of money is involved in the use of spies and hopes to make them accept this way willingly.

Sun Tzu says:

> Hostile armies may face each other for years, striving for the victory which is decided in a single day. This being so, to remain in ignorance of the enemy's condition simply because one grudges the outlay of a hundred ounces of silver in honours and emoluments, is the height of inhumanity. One who acts thus is no leader of men, no present help to his sovereign and no master of victory.

The Yokan Denkai says –
Koka traditions say:
The ideogram used in the quote above 仁 is read as 'Megumu' – which means to cherish. Those who have a selfish will and are miserly do not cherish. Therefore, they are forced to spend drastic amounts of money on military necessities when urgent situations arise, at the same time they choose not to spend coin on spies and do not utilise them, as they despise outlaying funds in this area. These [weak generals] do not understand the common saying 'trying to retain small amounts will result in a high expenditure'. This is why Sun Tzu emphasises this point in the above quotation.

Iga traditions say:
The greatest fault for a commander in chief is not being benevolent. Unless he is benevolent, he will not achieve great feats and none of his retainers will have admiration for him. If those retainers do not have admiration, then it will be difficult for him to achieve his goal. Especially, as he desires money and does not love his people, then he will not use spies and not be able to accomplish any great deeds. Sun Tzu warns against this.

The Yokan Rigen says:
'Hostile Armies May Face Each Other for Years.'
This refers to the situation where you are confronting an enemy face to face. It sometimes does not take many days to defeat them, sometimes it may take a month or sometimes it may even take a year to decide victory or defeat, however, also it can sometimes take years to gain victory. The battle of Okehazama, conducted by Lord Nobunaga, took only one day while the Mori clan's campaign against the Amago clan of Unshu province took seven years. When Shingen attacked Shinshu province, it took years too. There are numerous similar cases. [Sun Tzu] says 'years' with the intent to make it sound like a very difficult situation. Even after years of conflict, it will take only one day to decide an outcome in the end. In order to achieve victory on the day of battle, they may confront each other for years. If they are an army of a hundred thousand people, they spend one thousand pieces of gold every day, creating uproar at home and abroad, exhausting the high and low. If the general does not care for this fact, then he is an ignorant general. Even if he is victorious when following this way, the country will collapse, the people will be exhausted and disaster will arise around his feet. Therefore, if the general uses spies to closely know the enemy status and take appropriate measures and win, it will not take years or months and will cut down on expenditure, be it state or private spending, it will reduce the number of fatalities, gain an easy victory, release the people, enforce the province, and enrich the region. There is nothing else which can bring such great

benefits. What made the victory in Okehazama possible first of all was Lord Nobunaga's use of spies against the province of Suruga. He had them investigate their internal setup and discovered that a capable general from Owari whose name was Yamazaki Shinzaemon[23]was being entrusted with the very important position of command of the whole army. It seemed that [Imagawa Yoshimoto] would swallow up all Owari and even the whole country in the end if Yamazaki was left alive.[24] So Lord Nobunaga constructed a plan to kill him and found that the defender, Shinzaemon, had splendid handwriting. He collected such writings and had a very skilful secretary learn to copy his style for two years. After the two years, this agent became so good at copying Shinzaemon's handwriting that it was difficult to tell which was his true hand and which was a copy, then at this point Lord Nobunaga had him forge a letter implying a secret communication between Yamazaki Shinzaemon and the Oda clan. He had his trusted retainer, Mori Sanzaemon, disguise himself as a fishmonger from Atsuta and go on a trading mission to Sunpu. Mori Sanzaemon made an acquaintance of a close retainer of Imagawa Yoshimoto and had him spread rumours that the defending leader Yamazaki Shinzaemon was playing both sides and this rumour reached Yoshimoto's ears, and Mori arranged it so that Yoshimoto would see the forged letter via the retainer. The letter arrived after the rumour saying that Yamazaki Shinzaemon was playing both sides had emerged and the lord saw this letter, which looked like Shinzaemon's own handwriting; because of this Yamazaki Shinzaemon was killed immediately without any thorough investigation into the matter. After that there was no retainer who could match Yamazaki Shinzaemon in skill and serve in his place, so their tactics declined and this ended with Yoshimoto being killed at the battle of Okehazama.

All the above activities such as forging handwriting, making an acquaintance of a close retainer, and the tactics Mori used, these all cost money and excellent generals, ancient and modern, in both Japan and China, know this and offer shinobi no mono fief or official position liberally to cover such costs. If shinobi are satisfied and carry out their missions successfully, it is possible that when a battle takes place, your province will not be exhausted and you will be able to win without fighting or gain total victory with ease. On the other hand, if you do not care how many people – be it samurai or common – will be killed or injured and are willing to join in battle or stay in a position for a prolonged period, regardless of how much money will be spent, and if you oppress people by imposing tax or labour on them, then know that this is extremely unrighteous. Such a general should not govern people, nor assist his lord. If the lord is as unwise as the general, then they will not understand that using spies will gain them an easy victory.

[The ideograms used in Sun Tzu's work;] 所以動而勝人成功 mean 'to strike and conquer, and achieve things' and imply that before you dispatch your army to war, you should send spies to investigate the following closely: the topography of the enemy province, if the lord and if the people are in accord or discord, if the enemy army is strong or weak, if the politics are good or bad, if they have sufficient military defences or not and so on. Also, if you cannot manage to settle a situation without staining your swords with blood or finding a way out without fighting, only then, after knowing these things, should you make decisions according to the conclusion your council has reached and raise an army. If you check thoroughly any plan decided on against the actual situation presented on the day of battle, and make your move appropriately, then you will overcome if you attack and you will win battles and realise great achievements.

[The ideograms used in Sun Tzu's work;] 出于衆者先知也 'achieve things beyond the reach of ordinary men' means to know [the enemy situation] beforehand. If done, you can have total victory as mentioned before; this is because you are beyond ordinary men and have a far better talent and intelligence than others and you will obtain the needed information before an emergency arises and take the correct measures to deal with a situation.

Sun Tzu says:

> Thus, what enables the wise sovereign and the good general to strike and conquer and achieve things beyond the reach of ordinary men, is foreknowledge. Now this foreknowledge cannot be elicited from spirits; it cannot be obtained inductively from experience, nor by any deductive calculation. Knowledge of the enemy's dispositions can only be obtained from other men.

The Yokan Denkai says –
Koka traditions say:
When the god Susanoo confronted the giant snake, and when the god Yamatotakeru confronted a massive army, they won without difficulty though the situation was truly dangerous. They achieved great exploits with ease, as if their passage was unhindered. This is because they knew the situation of the enemy and they could 'see' this as clearly as reflections are seen on a clear mirror, and with the information they gained they won without hindrance. The essence of how to win in warfare lies only in knowing the enemy status, there is no other way to obtain the information than through the use of spies.

The Yokan Rigen says:
[The ideograms used in Sun Tzu's work]
先知者，不可取于鬼神，不可象于事，不可驗于度，必取于人，知敵之情者也 mean, 'Now this foreknowledge cannot be elicited from spirits; it cannot be obtained inductively from experience, nor by any deductive calculation. Knowledge of the enemy's dispositions can only be obtained from other men,' This shows that such wise lords or good generals as mentioned above do not pray to spirits, perform divination or fortune telling by divining sticks to obtain information. Nor do they obtain it by studying ancient writings or ancient battles, or by studying those cases or matters appearing in those materials. Also, they do not gain information by observing the art of In-Yo – the study of natural phenomena, be it wind or clouds, the degrees of the sun, the moon, and stars etc. Information is gained by nothing but the use of spies and they obtain their information before any situation arises. Wise lords and good generals have wisdom and are talented and they always keep spies in close service and have them investigate the manners and customs, the political dynamics of the lord and his retainers, if the samurai and common people are in accord or discord, the entertaining arts, children's songs of the enemy province, etc. This information concerning an enemy status will allow them to make proper decisions on how tactics should be carried out. If it is not achieved by the use of spies, then it is difficult for you to know of the enemy's status. Therefore, 'other men' in the quote above means spies.

The Koka traditions have a deep secret oral tradition and this tradition should not be transmitted lightly. The tradition is thus:

Good generals do not use spies only to obtain detailed information about the enemy but also they use them to thoroughly spy upon their own allies. By giving an example of an old incident, this point can be understood thoroughly.

In 1600 Ishida and Uesugi rose in rebellion, however, [Tokugawa Ieyasu] had a thorough knowledge of the situation as he had investigated everything beforehand, well before any wars had started, when the appearance of an uprising had not even surfaced, he had done this to decide which Daimyo would take his side in any wars to come and who would be faithful. With this information, [Ieyasu] decided on the number to be sent to attack Uesugi, and went east with them [to attack him]. Ishida took advantage of this chance and raised an army. Japan was thus divided into two factions, east and west. However, when Ishida raised his army, only Tamaru and Miyabe of the eastern forces took sides with Ishida's western forces. On the other side, Ishida did not know that the Mori and Kobayakawa clans were 'sitting on the fence' and that they had, in the end, decided to join Tokugawa and because he did not know this, Ishida relied heavily on them. But this misplaced trust ended in him being defeated. On the day of the battle of Sekigahara, if the Mori clan had joined together [with Ishida] and if Kobayakawa Hideaki had not betrayed [Ishida] and even though Lord Ieyasu was supremely excellent [as a general], it may have been the case that he may have lost. Ishida's side were uncertain as to whether they could trust Kobayakawa or not, but as for Mori, they were completely unaware of his betrayal. This situation arose because Ishida did not send spies to investigate his own allies. He did not send spies because he was over-confident about his abilities and therefore he did not pay much attention to the feelings of others. The following is to be kept secret, *Lord Ieyasu always used spies to know both the enemy and his allies*, which enabled him to win in all battles. More information can be found in the complete record of *The Battle of Sekigahara*. Refer to this writing for further consideration.

This is the end of the discussion about knowing the enemy status by use of spies.

THE USE OF THE FIVE TYPES OF SPY

Sun Tzu says:

The use of spies, of whom there are five classes:

1. Local spies 郷間
2. Inward spies 内間
3. Converted spies 反間
4. Doomed spies 死間
5. Surviving spies 生間

In this text (Yokan Denkai) I Shigenori use the term In-Kyoukan 郷間 for Local Spy, however, in other manuals[25] Local Spies are called In-kan 因間.[26]

Sun Tzu says:

> When these five kinds of spy are all at work, none can discover their secret systems. This is called 'divine manipulation of the threads'. It is the sovereign's most precious faculty.

The Yokan Denkai says –
Koka traditions say:
The path of the shinobi is marvellous and unforeseeable. This is because [shinobi] were originated by a god as is told [in the origin myths.] [The way of the shinobi] is called a divine marvel as nobody can outwit [shinobi] because their way is beyond description and there is no end to their tactics. In using spies: if [spies] are not good enough to perform divine or wondrous feats, then they will not achieve, while if they do perform to such a level, then they are truly the treasure of warfare, and nothing is truer than this fact. Unless the commander has accepted this point, all will be disastrous and will lead to a downfall in the end.

Iga traditions say:
There are various kinds of treasures, such as gold, silver and the seven jewels.[27] What are thought of as treasures is different according to the situation and the person. In the writing of Chu 楚書, Virtue is regarded as treasure, while in the Book of Rites 礼記[28] parents are regarded as treasure. In warfare, Sun Tzu says that spies are treasure, and he is the greatest person of all those who said this in Japan and China. We are in a land of different roots of gods or spirits [from those of China] and now few people think that performing the art of spies should be respected or valued as a *treasure* for a general, but I think this is wrong. The fact that Sun Tzu used the ideogram for *treasure* 寶 is a great gift indeed. Those who come after should accept and appreciate it.

The Yokan Rigen says:
Sun Tzu's phrase 'all at work' does not always mean to use the five types of spies all at once. It means to use one or two, but it can possibly mean all five [depending on the situation], but the number of forms of spy will depend on and be in accord with the situation. 'None can discover their secret system' means that wise lords or good generals use [spies] in such a subtle way that people do not understand how they do so. It is exactly the same as when bad *Go* players cannot tell why a good *Go* player makes a certain move. The ideograms for 'divine manipulation' are 神紀 and the second ideogram 紀 means 'rules' or 'laws' making the two together 'divine laws' so the text implies that you can know wondrous and indescribable things and that you will have the ability to win in every battle [with the use of spies]. Thus nothing is more precious a treasure [than spies].

THE USE OF LOCAL SPIES

Sun Tzu says:

Having Local Spies means employing the services of the inhabitants of a district.

The Yokan Denkai says –
Iga traditions say:
Sun Tzu's Five Types of Spy are numbered from one to five, with number one being the easiest. Local Spies are numbered as one – the simplest form – this is because this art ranges widely and is easy to perform.

[Iga] traditions also say:
In our traditions concerning shinobi, you should make solid connections with local people of neighbouring provinces and utilise them, do this during peacetime before a confrontation occurs between your province and the enemy. If you try to make a connection with peasants or merchants in haste after you are in a confrontation with that area, know that it will be difficult as they will have put in place strict guards along the boundaries and will have prohibited trespassing, making it difficult for you to enter freely. There is an oral tradition concerning *The Principle of Oshinobi* 大忍ノ大事.

Koka traditions say:
It is very easy and yet at the same time it is very difficult to obtain information about the enemy status by utilizing peasants or merchants. It can be very difficult because after a confrontation occurs, boundaries will be strictly guarded and the entrance of those travelling from other provinces will be strictly checked. Also it can be easy because when it is peacetime, you can go back and forth without problems or can even live in another province with ease. Furthermore, as it is difficult to venture out to distant provinces, you should at least make acquaintances in neighbouring provinces around your own state and remain in contact with [the people of the area]. By doing this and when needs arise, you can have your spies move into one of those provinces to carry out any plans required. This is called *The Principle of In-Yo Shinobi* and there are oral traditions for this art. The deep secrets that a leader of shinobi[29] should keep in mind lie here.

Ancient sources say:
Treat local people generously and use them; know the enemy's tactics by using locals; approach those of an enemy province generously and with politeness.

The Yokan Rigen says:
Local spies should make acquaintances with local peasants in the enemy province and use them discreetly. Even though they are only peasants, it is not easy to make them follow you and act as you wish in secret and also, know that there are thousands of various skills on how to make connections with them. Take note, if you seek for those peasants who are originally from an enemy province but who live in a different province or in your own province – which may be for various reasons – then convince them to make connections with others in the intended enemy province and through this connection, you should bribe those living in the enemy province as well as the one(s) from the enemy province [who you originally made the connection with]. Achieve this with gold, silver or cloth. You should provide benefits for them and angle the situation so that they feel grateful towards you, or overwhelm them by letting them know how powerful your province is, inspire them with awe by informing them of all the benefits that they will receive, also encourage them to have sympathy for you. Alternatively, there may be a ruined samurai hiding among the local peasants, so you should give him a position or fief and by giving him such an incentive, have him contact your spy in the enemy province, so that he can return to your side with information. No matter how strictly the enemy have guarded their boundaries at each and every checkpoint, if local people from the enemy province pass through it, then both your spies and the enemy spies can move freely.

THE USE OF INWARD SPIES

Sun Tzu says:
Having Inward Spies, making use of officials of the enemy.

The Yokan Denkai says –
Koka traditions say:
Although it seems to be difficult to make Inward Spies during peacetime, it is very easy during a warring period. You should know how easy this is by considering those lessons from ancient wars. People's disposition differs if it is a time of peace or a time of war.

Iga traditions say:
[Inward Spies] are ranked second because it is a more difficult task than Local Spies, however, it is comparatively easy, be it a time of peace or of war, but know that in some areas, this depends on the disposition of those in the clan. You should consider the customs and manners of the intended province when you conduct [Inward Spies]. The *Art of Fleas and Lice* [is connected to this and] is orally transmitted.

The Yokan Rigen says:
Here 'officials' 官人 means those who serve the enemy clan with a position and fief, they are generals and bugyo magistrates. [The following are those people you should approach:]

- Someone who is talented but who is not given an important position and is in social decline.
- Someone who failed and was deprived of his position, and has been restricted to his house.
- Someone from a family which has had a criminal executed.
- Someone who envies his colleagues for their successes and promotions.
- Someone who has a suitable position but is greedy.
- Someone who is conceited and likes disturbances.
- Someone who has no loyalty and is dishonest and will betray others without hesitation.

You should approach those above people as listed and: explain to them the differences between a righteous army and an unrighteous army, tell them about what is beneficial to them, impress them by displaying how powerful your forces are and overwhelm them, promise them position or reward, bribe them with money, treat them kindly and meticulously and convince them by way of careful speech and writing. Through these ways they will, in the end, give allegiance to your side. [After winning them over] you should have them bring a colleague of high rank with a large fief or even someone of lower rank with a small fief over to your side, or to cause a discord between the lord and his retainers and make them communicate secretly and betray your enemy. This is what the quotation 'making use of officials' refers to.

THE USE OF CONVERTED SPIES

Sun Tzu says:

On Converted Spies – getting hold of the enemy's spies and using them for our own purposes.

The Yokan Denkai says –
Iga traditions say:
There is no other way [that you can manipulate the enemy] than Converted Spies when they are strictly guarding their boundaries, it is difficult to have your spies infiltrate by any method in this situation. Even if you can successfully prevent enemy spies from infiltrating your position, it is still difficult to have your spies infiltrate theirs. Therefore, if enemy spies infiltrate, you should intentionally let them in and utilize them for your own purposes. *The Art of the Moon on the Water* is a tradition just for this purpose.

Koka traditions say:
You should investigate and know exactly how the enemy spy feels and make your plan based on meeting his desires with the aim of making a Converted Spy of him. If you do not know what he is thinking or how he feels you will not be able to convert them. You should try to find out the state of his mind with the *Traditions of Four Ways of Knowing*.

The Yokan Rigen says:
Converted Spies are enemy spies who have come to your province to investigate everything and report it to the enemy lord and also to mislead and disturb you by spreading false information. Wise lords and good generals will find them and pretend to be unaware of their presence with the intent of using them to realise their own plans, these spies are called Converted Spies. This art cannot be performed without being subtle. Enemy spies sometimes come [to your province] when needs arise or have been placed in your province as samurai or commoners beforehand. Therefore, even before an emergency arises, you should check the registered population of your local people at times and have Metsuke or Yokome inspectors find them. If they find someone suspicious, you should pretend to be unaware of their actions and arrange so that they can remain in the area with no problems. Then you should make a false plan or fabricate a story and let the enemy spy know of it so that he will report back to his own province. With this false information, the enemy general will take measures to deal with the 'situation'. As you have given away false information to the enemy spy, next you should carry out your real plan to attack them at the appropriate time, in this way enemy plans will fail and you can halt their advance, if they retreat, you can block them, trapping them so that they can neither advance nor retreat. You should take advantage of this and make an unexpected move, which is completely different from what the enemy shinobi have reported, this will gain you victory without fail.

THE USE OF DOOMED SPIES

Sun Tzu says:

> Having Doomed Spies, doing certain things openly for purposes of deception, and allowing our spies to know of them and report them to the enemy.

The The Yokan Denkai says –
Koka traditions say:
Sun Tzu mentions that you should provide enemy spies or your own spies with false information and let them inform the enemy of such matters. However, it is unknown if it is the case in foreign countries. In our country people are highly intellectual and they will not make important decisions based purely on information gathered from spies alone. Therefore, you should release a prisoner or a great offender to bring about the planted false information alongside the Doomed Spy, so that the enemy will be truly deceived. This is called *The Art of Secret Connection* and is performed by those in our country who know military tactics well. More is to be orally transmitted.

Iga traditions say:
[In this art] you should first give away [false information] to your spies or enemy spies and then release a prisoner or a great offender by 'mistake', so that they will provide the enemy with [false] information. This is an art Sun Tzu did not mention and has originated in our country. This is called *The Art of the Three-way Connection*. There are secret skills for this art. If the enemy have performed this art on you, you should use *The Tradition of Separation* in response and there are secrets to be transmitted on this art also. If it is difficult to use *The Tradition of Separation* and you cannot help but confront the enemy, you should use *The Secret Art of Correct Connection*. More details are to be orally transmitted.[30]

The Yokan Rigen says:
In Japan's warring period, generally, spies were concerned about secret plans and plotting but were also concerned with information about what was not of a serious nature, such as the manners and customs and social conditions.

Remember, the art of using spies is not easy, especially the art of Doomed Spies is very difficult and that of Converted Spies is the second to this. Therefore, there are various theories about Doomed Spies.

In the Kanrei manual, Master Saigyoku says:
If your general is going to use Doomed Spies, you should keep the original plan you actually are going to carry out secret and not give it away to anyone, including your family or close retainers. Then you should construct a fake plan and discuss it with your close retainers and allow it to spread to those retainers who are outside of the lord's direct retainers,[31] allow this to happen naturally. Though people know of this fake plan, you should pretend to endeavour to keep it secret and bring the fake plan into operation, this will allow your spies to know of the operations that have been put in place, information which will reach the enemy spies in the end. The enemy spies will want to have evidence of your plan, so you should condemn and insult one of your own spies so that he will lose face and form a grudge, then he will give away the entire false plan to the enemy spy.

The enemy spy, having heard this, will believe him and report it back to their commander in secret. To ensure the success of this even more, your general should let a prisoner or condemned convict see or hear of the information and let them escape by allowing restrictions on him to become lax. The prisoner or the convict will run to the enemy province and report what he has heard or seen, and the enemy lord will find that what the spy says and what the convict says happen to be the same, so he will become pleased and make a plan [to deal with the information.] But then, you should carry out the original plan and attack at the appropriate time so that you will attack where they do not expect and what they have prepared against you will not succeed. This will result in your complete victory and the enemy general will be enraged and kill the spy and the runaway offender. This is why they are called Doomed Spies.

Master Saigyoku also says:
Although the situation fits in with this text, how can it be possible that an enemy spy will believe you and believe that your own spy is telling the truth? If he does not believe the spy, then he will not report the misinformation to the enemy province, or if he does report it he may state that it is not true. People say that Sun Tzu should have written that both the spies would be killed in such a situation, [the Doomed Spy and the Enemy spy]. However, even if it is beneficial for your province and aids your plan, it is still not a benevolent act and it is an awkward subject. So it appears that Sun Tzu wrote this in an indirect way. Those who learn should be aware of this.

Iga traditions also say:
If it is the case that the enemy uses the art of Doomed Spies against you, you may fall into the enemy's trap [if you believe what the spy says] or it may be the case that you can hardly tell what is true and what is not but be tempted by the possibility that you may win a massive victory. Alternatively, if you have won by taking your spy's advice on the previous instance, you may easily believe what your spy says again, but again you may still not be sure if it was the case that the enemy lost intentionally at the last battle or if you actually won in truth. For these cases you should use *The Tradition of Separation* in response. There are secrets to be transmitted on this art also. If it is difficult to use *The Tradition of Separation* and you cannot help but confront the enemy, you should use *The Secret Art of Correct Connection*. More details are to be orally transmitted.

The details in the above skills are not mentioned by Sun Tzu but are secret traditions. Therefore, you should not give them away without care but keep them secret.

In [ancient China] in the reign of the State of Song, the army from Xi-xia area was so powerful that a retainer of Song, whose name was Captain Cao, plotted and freed a condemned convict making him a monk and shaving his head. The captain wrote a letter and sealed it with wax in a ball and then made the monk swallow it, telling him to go to Xi-xia and to give the wax sealed letter to such and such a person by passing it through his body and that he would be rewarded if he did so and also, that if he was suspected, there was no evidence to find on him, so no harm could come to him for this ruse.

Captain Cao released the convict and had a man follow him. The convict-monk went to Xi-xia and the guards suspected and questioned him. The questioning made them even more suspicious so they tortured him, and the monk confessed that he swallowed the letter. They forced a purgative on him and obtained the wax sealed letter and opened it. It turned out to be a [fake] secret letter from the emperor of Song to a close retainer of the king of Xi-xia about a secret plot. Hearing of this, the king of Xi-xia became enraged and killed the retainer and the monk in the end. This is how you can have an enemy retainer who is gifted with ingenuity killed by using the art of Doomed Spies.

Emperor Gaozu of Han was trying to kill the king of Qi but found it was difficult to do so because Qi was so powerful. Then the emperor sent an advisor whose name was Li Yiji and had him talk about the advantages and disadvantages and persuade them to accept reconciliation. When they agreed to surrender they relaxed their guard and while Li Yiji was still staying with them, a general of the Han side, Han Xin, suddenly attacked the land of Qi and defeated them, which made the king of Qi very angry and he killed Li Yiji immediately. This was also an example of Doomed Spies.

[In Japan] Kenshin invaded Ecchu province, where the Shiina clan, Jinbo clan and other clans were fighting against him – however, Kenshin was so powerful that they could not stop him. Meanwhile, having heard that Kenshin restrained himself from having relations with women but that he indulged himself with young boys, they found a boy who had true loyalty to their cause, and pretended to banish him. The boy went to Echigo and into the service of Kenshin with the intent to kill Kenshin by looking for an opportunity; however, Kenshin identified the plan and killed the boy. In this case, Shiina and Jinbo used the art of Doomed Spies, but it did not work on such a wise general as Kenshin.

You should consider these ancient incidents in Japan and China and learn the art of Doomed Spies thoroughly. [In the last example] the boy was only young and was a loyal and devoted retainer. Therefore, it was not likely that he gave away the plan and Keshin should have sent the boy back to Ecchu immediately. Killing this boy was a totally wrong reaction from Kenshin.

In the Yoshino Shui[32] collection of stories there was a warrior whose name was Uno and he was in the vanguard of Akamatsu's army and was killed in a battle [against Kusunoki Masanori]. Akamatsu intended to use Uno's son to kill Kusunoki Masanori. Masanori – who may or may not have known of this plan – took him in and educated and favoured him, but was very cautious and did not let his guard down so that the man's son could not find any chance to stab and kill him; and this went on for years. When the [seventh] anniversary of his father's death came, as he had matured and gained his strength, he hoped that he could kill Masanori, but Masanori suggested that he should celebrate his coming of age and appointed his retainer Wada as his godfather, who named him Wada Kojiro Masahiro. Next he was given armour by the southern emperor, which greatly moved him and thus he began crying and tried to kill himself [because of his impossible situation]. At this point he revealed his true intentions and became a priest by the name of Shokaku Hoshi. He wrote down the entire tale and sent it to Akamatsu, together with a sword that Akamatsu had given him. Considering the former example, it was easy to understand that Kenshin was not deceived by such a plot but it was not righteous or benevolent that he killed the boy.

THE USE OF SURVIVING SPIES

Sun Tzu says:
Surviving Spies – finally, are those who bring back news from the enemy's camp.

The Yokan Denkai says –
Koka traditions say:
[Being a Surviving Spy] is exactly what shinobi do. The other four kinds of spies should be made according to the situation when needs arise and do not hold permanent positions.[33] Surviving Spies should serve in a permanent role and should always keep improving and preparing themselves without negligence, not letting up for even a single day. You should not be negligent in training and study from morning until night. There are some secrets to be orally transmitted on how to enlighten your wisdom.[34]

Iga traditions say:
To be a Surviving Spy is very difficult and thus they are ranked as fifth and highest. Take note that, the way of the Surviving Spy is the definite way of the shinobi. The four other kinds of spies are only tactics to be performed to deal with a situation at hand. However, there are only a few spies who can perform this path [of Surviving Spy] properly. If a general can obtain such spies, they are to be considered a treasure to the general.

Also the traditions [of Iga] say:
It is strongly suggested that tens or hundreds of spies should be used. There is no one who is good at every art [of the shinobi] so therefore, you should use them according to the talent or ability of each. Furthermore, it is very important to choose a leader who can use them freely. If a leader knows much concerning [the management of spies], then those spies under his command will be able to fulfil their missions without fail. Oral traditions on *The Art of the Pivot* are transmitted on this subject.

Also the traditions [of Iga] say:
Perceived mysterious and phantasmagorical skills[35] such as climbing up to a height, passing under a low [obstruction], crossing over water or marshes, climbing a wall and infiltrating a castle, hiding themselves, erasing their shadows, etc., are skills which Surviving Spies should only perform to deal with emergencies and are not skills that are considered their main role. It is completely wrong to think that such skills are the

righteous path and are the [only] skills of shinobi and you should remember that shinobi [ways] are often misrepresented as evil arts and vicious plans and deceptive strategies. [Students of shinobi no jutsu] should first learn the concepts of the correct way [of the shinobi] and after mastering them, acquire such strange skills as mentioned above and be flexible according to the situation.[36]

The Yokan Rigen says:
Surviving spies are those spies you send to the enemy provinces and who investigate if the enemy are in advance or in retreat, the movements of the enemy, if something is substantial or insubstantial, etc., and they report this back to your side discreetly. 'Bring back news' from the above quote means to report something while they are staying in the enemy province. Therefore, unless you choose an appropriate man after judging his character thoroughly, the result will be harmful for your cause. You should choose someone who is honest and faithful, not greedy, gentle, capable, wise, fluent in speech, patient, most courageous, and excellent in various arts. You should send him to the enemy province and have him live there and serve or make acquaintances with those of high position. Sometimes, you should have him disguise as a Shinto priest, Yamabushi, monk, merchant, doctor or entertainer, or making him a messenger is also an option. This will all depend on the situation. As mentioned above, those who investigate the enemy manners and customs, politics, and tactics and report the information back to your province are called Surviving Spies.

Some say:
If you pay attention to the ideogram 反 'back', you will understand that Surviving Spies serve in the enemy province and send the information 'back' to their home province. These spies spy and report back while they have a position within the enemy, and this is what is meant by [Sun Tzu] saying, 'bring back news'. There are a lot of cases of this in China and our country as well. We can see this in a relatively recent case, at the Siege of Osaka Castle, Obata Kanbei, being on the inside of Osaka Castle, sent reports to Itakura Iga-no-kami in Kyoto.

Statue of Sun Tzu in Yunhama, Tottori, Japan.

NOTES

1 That is, do not think that amazing and clever skills are what make up a good shinobi, it is a proper understanding of the essence of the use of spies and the principles that are found within Sun Tzu's chapter that are of primary importance .
2 The 'feeling' from the text is that Master Yorihide was not from Iga himself but that he was connected to Iga and their skills.
3 22nd or 23rd day of the lunar month.
4 As stated in the introduction, this section of the manual has been drastically shortened and condensed as notes.
5 Most likely, before the coming of age ceremony, therefore for those below and between the ages of 15 to 20.
6 Chikamatsu's name does not appear in the text and this manual was written three years before he died, but appears to be in the same handwriting and is considered one of his manuals.
7 Translated in full earlier.
8 As mentioned previously, the pheasant has no name or is unnamed and that it not respectable for the origin of the shinobi arts.
9 The Princess.
10 A mythological character, a kami of the land who appears in the 'Nihonshoki' (Chronicles of Japan).
11 Another mythological character.
12 The thirteenth emperor in the 'Age of Myths'.
13 A third mythological character, a military leader who according to some accounts was defeated by the first legendary emperor Jinmu.
14 The term here is Heiho and means military tactics and is not in Sun Tzu's book.
15 The quotation has been removed.
16 夜盗組 'night thief-group'.
17 This version implies thief.
18 Seeing and listening, discussed in depth elsewhere.
19 Heiho, the arts of war, not in Sun Tzu's text.
20 It is not clear if this means two kinds of people or two types of tasks.
21 The main point here is that an army on the march is expensive and the overall distance and time period is an unknown but expensive factor.
22 In the era of peace, lords had to journey to the capital of Edo on a regular basis; this was done to deplete them of funds.
23 Also known as Tobe Shinzaemon.
24 At the battle of Okehazama, Oda Nobunaga arranged a raid on Imagawa's army and took the head of Yoshimoto in 1560. With this victory, Nobunaga took the first step to becoming the ruler of Japan.
25 講義-開宗-集註
26 This sentence has been edited.
27 Gold, silver, lapis lazuli, corallite, agate, pearl and amber. The original text states: 'gold, silver and the seven jewels' making a total of nine, however gold and silver are included within the seven jewels.
28 One of the Confucian classics.
29 忍ノ総宰タル者 shinobi no sousai taru mono.
30 All of these arts are translated earlier in this book.
31 Tozama.
32 A collection of stories about the southern court of Japan.
33 Kimura is saying that this form of spy is the fully paid, fully trained agent and is not a temporary measure, what we would consider a professional shinobi.
34 The oral tradition states that you should learn from those families of Iga and Koka.
35 The text is saying that people perceive such skills in this way, not that the skills themselves fit such a description.
36 A shinobi should learn the fundamental principles of spying before they move on to such skills as those listed.

FURTHER TEACHINGS FROM SUN TZU

Sun Tzu says:

> None in the whole army have more intimate relations to be maintained than those relationships with spies. None should be more liberally rewarded. In no other business should greater secrecy be preserved.

The Yokan Denkai says –
Iga traditions say:
If you do not show affection to someone, you cannot know what the person thinks inside of their mind and nor can you have a close relationship with that person. Therefore, you should try to have as good relations as possible and to get to know the person as much as possible to be able to utilise them fully. Unless you have a close relationship, like that of father and son, they will become the enemy's Converted Spy. The way to prevent [your spies] from being converted is to act as a father does, as said above. There are oral traditions on this; *The Art of Connecting with the Hearts [of Other Shinobi]* and also *The Tradition of No Sword* are orally transmitted.

Koka traditions say:
The reward mentioned in the above Sun Tzu quotation does not mean position but instead means gold and silver. Generally, today it is difficult to use very noble and high positioned officials as spies, so lower officials or low ranking people are used instead. Secret skills about using them are concerned with gold and silver and this is why *The Art of Double Sowing* is used to prevent people from being Converted Spies.

Also the traditions [of Koka] say:
For generations, excellent generals have spared no expense where spies are concerned, nor do they ask for monies to be accounted for. On the other hand, ignorant generals who are ruinous to their own domains think the expenses for spies to be of no benefit and if money is given, they check how their spies have used it. In such cases as this, spies are not encouraged, and even worse, they may be won over and become the enemy's Converted Spies. Therefore, when you use spies, you should be as generous and think nothing of money, as if [gold] were mere stones. There are oral traditions called 'Points on the Grass Bundle'.

Iga traditions say:
Secrecy is primary in warfare. It is even more so for spies, since the matters that they deal with are extremely confidential and should not be given away, even between father and son, or husband and wife. If a plan is leaked, the army will lose any advantages that they may have. The key to keeping [a matter] secret lies in the two realms of close relations and reward.

The Yokan Rigen says:
It says 'intimate' because generally in war everyone has to devote his life to the cause, and spies especially are in danger as they are alone in the middle of the enemy, for this reason a general should try to have as intimate a relationship as possible and treat them with sympathy so that they will be loyal and true-minded to the lord. You should not leave it to other people to make them faithful because it will be difficult to communicate with your spies exactly your intended meaning and it will be difficult to understand what they want to report. So you should tell them what to do directly and make them understand exactly what you want and bond them to you. Therefore, you should have intimate relations with them, building bonds both substantially and emotionally.

It says 'liberally' because as spies are determined to sacrifice their lives regardless of how dangerous the situation is, they go to the enemy's land and fulfil their lord's mission, so they should be rewarded as liberally as possible. You should not spare reward, gold or cloth, and if you begrudge rewards, they will distance themselves emotionally from you and the faith of their principles will weaken and greed may grow in their mind so that they may be tempted by the enemy's bribes. Therefore, you should not be sparing of, reward and give them as many expenses as they need, thus [Sun Tzu] says 'liberally'.

[Sun Tzu] says 'secrecy' because with the secret mission given from the commander, spies venture to the enemy province and do not give away a single point but keep secrets only in their mind and report any information they have back to the general. If the spies are detected, they will be immediately captured and killed. So spies should be given their orders directly from the mouth of their lord into their own ears. This way, things are kept secret and they investigate the enemy status and report and nothing is given away to others. Nothing else is more secret than this – therefore, it says 'secrecy'.

Sun Tzu says:

> Spies cannot be usefully employed without a certain intuitive wisdom. They cannot be properly managed without benevolence and straightforwardness. Without subtle ingenuity of mind, one cannot make certain of the truth of their reports. Be subtle! Be subtle! And use your spies for every kind of business.

The Yokan Denkai says –
Koka traditions say:
Those generals who are not benevolent are always miserly. They do not spend a large amount of money so do not use spies. Refer to the previous chapter about the unscrupulousness [of those generals].

Iga traditions say:
If a person is completely endowed with these following three areas – intuitive wisdom, benevolence and subtlety – then not only can he control human affairs but also even the sound of wind or water will help his spies [to fulfil their aims]. 'Be subtle' as mentioned suggests that a lord should highly promote and entrust spies.

The Yokan Rigen says:
Spies cannot be usefully employed without a certain level of 'intuitive wisdom'. Generally, those who lead an army should be extraordinary in order to achieve victory. They should

know everything about the enemy such as the topography, the manners and customs, the political status of the lord and the retainers, and so on before a time of war comes. Thus they should use spies to investigate and know the enemy military status and tactics so that they can work out and construct a well-designed plan with a far-sighted and close viewpoint. Thus, they will succeed in every battle they fight and this is why talented and sage generals use spies.

It says that spies cannot be properly managed without benevolence and straightforwardness because, as spies go into the jaws of death to investigate the enemy status, their difficulty is unimaginably enormous. Therefore, [lords and generals] should have intimate relations with them, give them a lot of benefits, understand how faithful they are, have sympathy for their true mind and give them enough fief or a high position, so that a spy will be emotionally attached, bonded and devote themselves to the bitter end. If they are this firm and loyal, they will be willing to die and fulfil their mission in enemy lands without being tempted by any benefits or fiefs offered to them. Thus, [lords and generals] cannot use spies very well unless they are benevolent.

[Sun Tzu] says that without a subtle ingenuity of mind, one cannot be certain of the truth of a spy's report. In the word 'subtle' 微妙, the first ideogram 微 means something minute, hidden or that which cannot be seen. The second ideogram 妙 means 'wondrous and unforeseeable' and implies 'amazing beyond words'. The ideogram for truth 實 means the true status of the information from the spy. Spies investigate information about the enemy status, major or minor, not only the politics of the lord and the retainers and secret tactics but also the manners and customs of the high and low ranking people, the pottery they value, how they deal with messengers, rumours among local people, etc. The spy reports these things to his lord and the lord will contemplate the information, judging what can be seen and knowing which elements are hidden, and also speculate on what is hidden by knowing what can be seen, this is done to know the true status of the enemy. Thus 'make certain of the truth of their reports' means to fathom the true status of the enemy [from the compiled reports]. If you do not have intelligence enough to know of subtle things, you will not be able to get the truth from the information. If you are an adept at this, you will notice an enemy plot and not be deceived by them even if your spy is turned and becomes a Converted Spy. Thus it says, 'Be subtle! Be subtle! And use your spies for every kind of business' this quotation has the same meaning as: 'Spies cannot be usefully employed without a certain intuitive wisdom.'

Wise generals consider everything without exception, be it major or minor matters and they do this with subtlety so that they will know everything about the enemy status by using spies.

Sun Tzu says:

> If a secret piece of news is divulged by a spy before the time is ripe, he must be put to death together with the man to whom the secret was told.

The Yokan Denkai says –

Iga traditions say:

If various officials or bureaucrats fail to fulfil their jobs or leak a secret matter, they are rarely to atone for it with their death. On the other hand, spies are sentenced to death in an instant if they give away details of their mission. This is because if anything from their mission is given away, it can cause such a serious situation, such as an army being defeated or the lord and his clan being ruined. Therefore, spies are charged with extremely important and responsible tasks. It is wrong that there are not so many generals who understand this art very well and most of them think the task of spies is that of a low and humble position and do not have affection for their own spies.

Koka traditions say:
Of the three following things; affection, reward and secrecy – secrecy is of the utmost extreme importance. Although you should try to keep secrecy by way of affection and reward, there should still be the death penalty, to make sure that a secret will be strictly kept. By this measure, everyone will keep all secrets with care. There is *The Tradition of Securing the Source by Destroying the Ends* – to be orally transmitted.

The Yokan Rigen says:
'Before the time is ripe' means if any secret is leaked from a source that is not the spy before the spy has even been sent to the enemy province to start to send his reports back to the lord, then it may be the case that the spy has let [his mission] slip out. Or he may have written [to someone outside of the people who know of the operation] before he returns back to his own province, or he may have divulged information to someone he should not have before he reported it to the lord. In any case, if any secret is heard of from a source outside of the people that should have this information, know that it must have come from the spy at some point. If you track down the origin of the news by investigation and by talking to those who heard the information, there will be a source and a base point for the rumour. If that person says that he heard the information from the spy, both the man and the spy should be executed, this is done so other spies will never give away their secrets.

Sun Tzu says:

> Whether the object be to crush an army, to storm a city, or to assassinate an individual, it is always necessary to begin by finding out the names of the attendants, the aides-de-camp, door-keepers and sentries of the general in command. Our spies must be commissioned to ascertain these facts.

The Yokan Denkai says –
Iga traditions say:
The above is what spies should do above all and with a pressing need. You cannot achieve anything if your army goes to battle without knowing these points. Thus, it is without question that spies are the most important aspect and most urgent of all factors found within an army.

Koka traditions say:
It is absolutely wrong that – as other [schools] say – only vicious plans or deceptive tactics are essential in the use of spies. Sun Tzu suggests here that the arts of spies are not about deception or fraud but a universal way of all warfare.

Iga traditions say:
In the first chapter [of *The Art of War*], Sun Tzu talks about laying plans and discusses seven points[1] you should know [to determine the enemy's condition]. Thus the question,

how can you get information on these seven points? It is only possible through the use of spies and therefore their employment is of extreme importance. People know how important these seven points are but do not accept the truth that spies are essential as a foundation and nor do they have faith in the value of them. This is all because they do not know military ways in depth, they should study this thoroughly.

The Yokan Rigen says:
If you ever have an enemy army to attack, an enemy castle to capture or an enemy general to kill, before carrying out an intended plan, you should first know the names of the lord, the commander of the army, the commander of the camp, the other commanders, commissioners and captains, the lord's close retainers, the envoys, the captain of gate guards, the commander of the castle watch, and other officials. Then you should give these names to spies and have those spies get further information about each of them, such as: how wise, brave, talented, smart or ignorant, strong or weak they are, etc., and to make plans based on the information.

Sun Tzu says:

> The enemy's spies who have come to spy on us must be sought out, tempted with bribes, led away and comfortably housed.

The Yokan Denkai says –
Koka traditions say:
The first ideogram in the above quote is 必 'must' and is critically important here and should be analysed intensely. This ideogram is used earlier in a previous paragraph [which stated that knowledge of the enemy's dispositions can only be obtained from other men.] In both sentences it means the same thing ['without exception']. That information which is about people can only obtained by other people by way of asking others. [The art of shinobi] is not mystical and does not use unreasonable skills or the art of demons or divination.

Iga traditions say:
[Information] can be only obtained by people by way of asking others and that is why the ideogram 必 is used for emphasis. Because of the possibility of [spies] becoming Converted Spies, traffic between spies has not been prohibited.[2]

Koka traditions say:
Although it seems to be extremely difficult to use enemy spies for your benefit, you will be able to use them freely by bribing and housing them. Converted Spies can be used if you pay them, remember gold makes the use of your own spies much more fluid and easy.

The Yokan Rigen Says:
Generally, in the warring states period, the enemy infiltrated spies into your army to investigate your situation. If you detect them, taking care not to stir things up, you should have a tactician approach the spy and entreat him and allow him to remain without concern. You should offer him quarters so that he can remain [in comfort]. 'Tempt him with bribes' means not only financial benefits but also offering other benefits so that he

can stay comfortably. Also you should heavily bribe him with gold, silver, cloth, etc., do this to lure him in and make him forget his obligations and to sway his determination. Even if you cannot convert them, you can at least speculate on their tactics from the way they respond [to your questions.] Or let him know how powerful your province is, or feed him fake plans so that he will inform his side with false information and his province will then make wrong decisions based on the information. Therefore, you should find out the enemy spies and use them to aid your tactics.

Sun Tzu says:

Thus, they will become Converted Spies and available for our service.

The Yokan Denkai says –
Koka traditions say:
You may think it is difficult for enemy spies to infiltrate, but it is actually not difficult at all. Also you may think it is difficult for them to stay for a prolonged period, but it is actually easy [for them to remain near you]. During this stay, there will come a time when they are negligent and are found to be off guard. You should take advantage of this moment and bribe them to make Converted Spies of them. As soon as they let their guard down, it will offer a chance to sway their mind so that you will be able to obtain and use them as Converted Spies.

The Yokan Rigen Says:
As mentioned above, you should let the enemy spy stay, bribe them, or impress fear upon him by telling him how powerful your domain is or suggesting punishments that you will perform upon him; all of this is done in order to make him a Converted Spy.

Sun Tzu says:

It is through the information brought by the Converted Spy that we are able to acquire and employ Local Spies and Inward Spies.

The Yokan Denkai says –
Koka traditions say:
It is very difficult for your spies to acquire and employ Local Spies and Inward Spies after they manage to infiltrate. Opposed to this, it is very easy to do this by use of enemy spies who have infiltrated your side. There is nothing as clever as this strategy by Sun Tzu.

Iga traditions say:
In the above [Sun Tzu] quote, the ideograms 因是are used[3] and the text talks about how to obtain information from the enemy through Converted Spies. By getting information about the enemy status, within and without, with this you will be able to create, plan and use the [other] four types of spies.

The Yokan Rigen Says:
You should bribe the enemy spy to make a Converted Spy and have him make contacts among the local people and the officials. You then should make them feel indebted to you, impress them with your power or threaten them with punishments, promise them a position or reward so that you can use them as Local Spies or Inward Spies.

Master Saigyoku says:
Various annotations [on *The Art of War*] only take the above quote to mean luring the enemy spy to your side, making them change sides, betraying their own allies and working under you. However, if you use the enemy spy to know the enemy status, let him send information [back to his own people] about you to benefit your tactics and to allow you to fulfil your aim, it is the same principle as Converted Spies but without the need for him to change sides or betray his allies.

Sun Tzu says:

> It is owing to information [gained by a Converted Spy], that we can cause the Doomed Spy to carry false tidings to the enemy.

The Yokan Denkai says –
Iga traditions say:
After the use of a Doomed Spy has been successful and the enemy has suffered a defeat and the spy has been executed, the enemy army will become suspicious and hesitant in every battle when they take decisive measures. On top of this, [enemy] spies themselves will also have doubts and become troubled and hesitant and they will not divulge anything of what they can infer [from any given situation] or what they know by speculation. This after effect will benefit you ten times as much as the last victory you gained [from the Doomed Spy alone]. Therefore, once you succeed with a Doomed Spy, you will secure the foundation for an overwhelming victory. More is to be orally transmitted.

Koka traditions say:
In our country, there have been a vast number of cases where a victory was achieved by using Converted Spies but only a few cases where a significant victory was obtained by using a Doomed Spy. This is because the strategy of Doomed Spies has not been studied closely and mastered. More details are to be orally transmitted.

Sun Tzu says:

> Lastly, it is by using [Converted Spies'] information that the Surviving Spy can be used on appointed occasions.

The Yokan Denkai says –
Iga traditions say:
In the quotation, the ideogram 期 means '[good] occasion' and here 'occasion' should be considered as correct timing. Surviving Spies act on information from the Converted Spy, thus, if you do not have Converted Spies, your timing may be incorrect, however the use of Converted Spies brings correct timing.[4]

Koka traditions say:
The other four types of spies should be performed on a foundation of the information from the Converted Spy. The ideogram 因 from the above quotation means: 'by way of' or 'stick to'. More is to be orally transmitted.

Sun Tzu says:

The end and aim of spying in all its five varieties is to gain knowledge of the enemy; and this knowledge can only be derived, in the first instance, from the Converted Spy. Hence it is essential that the Converted Spy be treated liberally.

The Yokan Denkai says –
Iga traditions say:
As mentioned earlier, knowledge of the enemy's dispositions can only be obtained from other men, and this is mostly done through a Converted Spy. There is no other [main] source than this, both for the enemy and for you. Therefore, this matter holds true, both for you and your enemy.

Koka traditions say:
Remember, the ideogram 厚 from the above quotation is used with the same meaning as in the following quote, 'None should be more *liberally* rewarded.' You should provide both your spies and [converted] enemy spies with high reward. Even if your spies are the most loyal and faithful in the extreme, even if they are one in a million and even if they are willing to go through fire and water for their lord, even if they can never be bribed, no matter how high the reward offered to them, even then, you should reward them fully, thus they will be moved by this generosity and they will more than ever endure the unendurable.

Iga traditions say:
People in the world say that [the art of] spies is difficult to perform and they call it a strange art, an existence of transformation and mystery. This is because they do not understand the chapter of 'The Use of Spies' intimately. Is there any mystery or difficulty in people using other people? However, that being said, if you try to achieve [in this art] without using Converted Spies, then it is [difficult], as if you were trying to fly without wings. It is only through Converted Spies that this art is possible.

Koka traditions say:
It is easy to make a Converted Spy of the enemy spy, while it is extremely difficult to prevent your spies from becoming Converted Spies for the enemy. You should consider this in depth. Remember, Sun Tzu uses the ideogram 厚 and in this context means thousands or tens of thousands [of gold coins]; consider how deep [Sun Tzu] is.

The Yokan Rigen Says:
Gokan, 五間 the five types of spy means: Converted Spies, Local Spies, Inward Spies, Doomed Spies and Surviving Spies.

If an enemy spy comes to your side, you should treat him liberally and bribe him with gold, silver or fief generously to win over his mind. Or you should tempt him by showing how powerful your forces are or threaten him with punishments so that he will obey you. If he is single-minded and so determined that he will not follow you, you can [still] obtain information about the enemy status from the responses he gives to you. Therefore, there is no better way than through Converted Spies to acquire information about the enemy status. You should not be sparing with money or reward but bribe them liberally. When you send your spies into the enemy, they will be bribed with gold or silver, or beautiful women or threatened with punishments to shake their determination, or persuaded with speech – they may be lured with greed, fear of being executed and his determination may be swayed and weakened in time, no matter how firm his resolve. Even if he does not betray you, he may be led to give away your status through questioning. Therefore, you should choose one who is faithful, has lofty ideals and who is brave and loyal when you send them as a spy to the enemy. Considering the first part of this article, know that at times it is not easy to make a Converted Spy of an enemy, so you should be very careful in this matter and take various measures to do so.

Sun Tzu says:

> Of old, the rise of the Yin dynasty was due to Yi Zhi who had served under the Xia. Likewise, the rise of the Zhou dynasty was due to Lu Ya who had served under the Yin.

The Yokan Denkai says –
Iga traditions say:
This paragraph is an example of the statement that [knowledge of the enemy's disposition] can only be obtained from other men but not through any other skills. You should not be misled by various worthless interpretations [of Sun Tzu's text].

Koka traditions say:
In more recent years, spies have been assigned for only one art or one skill [and are not utilized in various ways as was done in the past]. This is because the chapter 'The Use of Spies' by Sun Tzu is not studied in depth and this above quotation is not understood or absorbed.

The Yokan Rigen Says:
Those who study warfare say that spies deal with intelligence between the enemy and their own allied forces. The primary aim for spies is to know the topography of the enemy province, if the mountains and the rivers are steep or not, the names of the high ranking people, how the enemy lord is controlling his army, if he is wise or ignorant, their tactics, the manners and customs, the emotions of the samurai and the common people, if they are in accord or discord, if they are many or not, strong or weak, trained or no and so on. On top of this, their secondary aim is to cause a rupture among the enemy, separate them from each other, mislead them so that they make wrong plans or tactics, cause discord between the lord and his retainers, make false charges against a loyal retainer or faithful warrior with the purpose of removing them, etc. If you use these arts for your own greed, they will be the arts of bandits. If being used for an army fighting for justice, they will be righteous skills. Both the primary and the secondary aims should both be used when needed.

Sun Tzu says:

> Hence it is only the enlightened ruler and the wise general who will use the highest intelligence of the army for purposes of spying and thereby they achieve great results. Spies are a most important element in war, because on them depends an army's ability to move.

The Yokan Denkai says –
Iga traditions say:
After all, the art of spies is about [obtaining information] and only from other men and is also to outwit others with your own wisdom. Therefore, those who are very intelligent perform this art well, while those who have only low intelligence do not perform it well at all. In short, high intelligence controls low intelligence, victory for the spy lies in this point. In the quote, the ideogram 上 'high' is considered most important.

Koka traditions:
For actions or movement without fear, people have to see with their eyes and know with their mind so that they can move around, stand up or sit down, deal with people, diligently study and train hard. All these things cannot be done without the ears and the eyes. Likewise, an army cannot move without ears and eyes. The ears and the eyes represent – as mentioned earlier – nothing else but spies. Just as people always move, dependent on their ears and eyes, an army can only move dependent on spies. There is no difference between these two points. How vast the use of spies is!

The Yokan Rigen says:
King Tang [of China] used a man of supreme intelligence as a spy and he knew the developments of his enemy and their strategy, whether it was substantial or insubstantial and through him they decided if they should advance or retreat or if they should carry out their strategy. By doing this, when they put their plan into action, they achieved great feats without exception. In previous quotations, the text says 'the wise sovereign and the good general, to strike and conquer, and achieve things beyond the reach of ordinary men, use foreknowledge.' This means that wise lords or good generals do not obtain foreknowledge by praying to spirits, performing divination or fortune telling to obtain information. Nor do they know things by studying ancient writings or ancient battles, or by studying those cases or matters appearing in those materials. Also they do not discover things by observing In-Yo natural phenomena, wind or clouds, the degrees of the sun, the moon, and stars. They obtain information on the enemy status only through men – that is those who are wise and intelligent, as explained earlier. With this being said, the conclusion here is that you should make a spy of one who is of supreme intelligence so that you will achieve great exploits.

It is essential in warfare to make spies of those who are of supreme intelligence as mentioned above. You should use them to gain intelligence on the enemy status, manoeuvre your army, exploit momentum, advance or retreat, move or stop, so that you will always be victorious. Sun Tzu establishes that it is spies that manifest and manoeuvre the forces of an entire army.

This is the end of the Yokan Rigen manual.

POSTSCRIPT TO THE YOKAN DENKAI [BY CHIKAMATSU]

Master Saigyoku-ken [of Naganuma Ryu] says that the use of spies, secret plans or plots, coups, and endangerment easily become vicious and evil stratagems so they are not what a lord should utilise. However, if a righteous army is going to be driven into a difficult situation and your ruin is imminent and if you have almost been gripped by the fangs and claws of a tiger or a wolf, then such a situation is inevitable. The above should be used to deal with the dangerous and urgent situations and to save virtuous people. An example of the above [skills]: if an epidemic is ranging to a level that is mysterious and unprecedented and strange, and if no medicine can cure it, then it is inevitable that a cruel medicine or extreme concoction or poison may be used to save people from death. After this, people will recover their true and original chi at a later date; this is the skill of a good doctor. These skills are not for ostentatious display but are an art which is benevolent and has to be done out of necessity. In warfare, being complete is the key to the victory. If you hate to employ anything that includes cruelty and atrocities and if you ignore the art [of spies], then, it will turn out that you will create a gap [in your defence], a gap which will be taken advantage of by evil doers. Furthermore, if you know of the art [of spies], you will cause your guards to become stricter and leave no opportunity for evil doers to infiltrate or invade. For this reason, Sun Tzu discussed this art [of using spies] in the final chapter to complement what other skills do not cover.

I Shigenori, say that Sun Tzu gives a warning every time he talks, and also so does our teacher, Master Saigyoku [of Naganuma Ryu]. Those who come after me should understand this and reflect upon it with deep consideration and respectfully and also, they should not be negligent. I would like to say this here, at the end of this book.

NOTES

1 (1) Which of the two sovereigns is imbued with the moral law? (2) Which of the two generals has most ability? (3) With whom lie the advantages derived from Heaven and Earth? (4) On which side is discipline most rigorously enforced? (5) Which army is stronger? (6) On which side are officers and men more highly trained? (7) In which army is there the greater constancy both in reward and punishment?

2 It appears that spies were allowed to communicate owing to the chance of gaining information and to convert enemy agents.

3 The ideograms are basic and mean 'use this', thus the point of the article is that 'this' in this context is Converted Spy.

4 The meaning of this sentence has been paraphrased from information inferred from the original text.

甲賀忍之傳未来記

Koka Shinobi no Den Miraiki

FOR THE PROSPERITY AND FUTURE OF KOKA SHINOBI

The following writing is the oral transmission of Sensei Kimura Okunosuke Yasutaka [of Koka] and is given to and written down by his student Chikamatsu Hikonoshin Shigenori.

Those who are given the traditions of the Koka shinobi arts should all understand the concept of 'Sansei' or the three stages of 'past, present and after', before they enter into study in this school.

- Past – that which has gone before
- Present – that which is now
- After – that which will come

In ancient times those originally from Koka used to consider themselves as one people and that they were all of the same origin, this was their standpoint for generation after generation and they treated each other like brothers from the same family. This even continued after they became scattered across the provinces and lived in various places, or in faraway territories, even then, they continued to know each other's names, and kept in contact constantly. Also, if need arose they communicated without hiding their intent and displayed everything [about the situation] and arranged to help each other to perfectly fulfil any shinobi mission at hand, and they did this without fail. They also made it a rule not to pass down any of their traditions to anyone who was not from Koka and this formed the core and guiding principle of their ways [of inheriting their arts]. Even today [in 1719] the above two[1] major principles are still observed and those Daimyo lords who want to hire shinobi no mono prefer to hire those of Koka.

It is unlikely that any warrior from Koka cannot find employment and that even if he is retained for only a small fief, he can easily gain employment. In fact, there are plenty of people [of Koka] who are given fiefs even whilst they themselves remain stationed in their homeland. Also, people think that simply being from Koka makes a person an expert in the shinobi ways, and this thought has arisen because Koka is run on two old traditions [and has its foundation in these] strong principles. Until now, this principle [of a strong union] has been fulfilled and those [of Koka] have let each other know who is in which province and also, those who are in important or larger provinces keep in contact more often than others and if an emergency arises, they will resolve any issue very quickly, so the reputation of the shinobi of Koka has been maintained and is renowned, therefore, for this reason they are hired by any and all clans.

As I am old, I am now considering the status of those who now do this same task [as I once did] and are spread out across various provinces and thus I am deliberating on the future prospects for our ways. Therefore, I will now make clear the present status and situation in detail, concerning our future[2] and I surmise that these following predictions will turn out to be the case in one out of ten cases. Also, I hope that my student writes down exactly what I say and shows it to all future students.

[The following will be the cause of the demise of the arts of Koka.]

I

Those people who instantly understand the importance and way of these two ancient principles and our traditions are those who lived within a short period after the turbulent times[3] and they have seen much, which has allowed them to fully understand their family arts 家業 without conscious study. And if you do not follow the way of these ancient traditions, then you cannot fully serve in an emergency and cannot fulfil the requirements of the appointed shinobi tasks of your family.

Those people around the age of 70 years old think of keeping these traditions alive as an important rule and still write to each other and do not break their promises which were made to their ancestors, sparing no effort in preparation, so that in an emergency situation and if the need arises they can be ready immediately the next day and therefore, those around 70 years of age will serve the clan 家 well. Those who are younger than 70 years cannot serve [their clans] very well and this should be a point for future discussion. The reason for their uselessness is that it is a time of peace in this land and there have been no crises during which those people from Koka can achieve greatness and show the arts of their family. Thus, because of this they have no opportunities and receive no rewards or recognition.

II

As there are no emergencies then there is no chance for our men to get close to their own lords and receive his orders directly or even converse with him. In olden days [Sengoku Period] there was the saying 'from the mouth to the ear' and Koka no mono used to talk to their lord directly and receive his orders and thus they built a good relationship, however, as this is not the case now, we have become lower in position and are of the level of Geshoku.

III

Because of this lack of emergencies, Koka no mono cannot achieve greatness nor accomplish feats, and there will be scarcely any lords who will want to use Koka people or appreciate them in the future and they will have no chance to prove their worth.

IV

As there is no demand to serve a lord or conduct official business, [new shinobi] will not maximise their own creativity.

V

As I mentioned previously, it is now a time of peace and so [the Koka no mono] will become accustomed to an easier life and will become negligent and have zero inspiration or will to expand their arts.

VI

[Koka no mono] are spread across all provinces but are now generations apart. In older days they used to be father and son or even brothers who were spread about the land, but now they are almost unconnected by blood and are strangers to each other, like those other people around them; and because of this they seldom write or communicate with each other.

VII

In times of peace, people do not remember times of war and what it is like and become comfortable, so they feel that there is no need for the shinobi and they no longer carry detailed information about each province, nor do they give gifts to each other. On top of this, they do not know who is in which area or who good people are.

VIII

Today, in every clan [shinobi] are not needed so they are left 'left to hang'[4] and many are given small fiefs and at a relatively low level [to how it used to be]. Therefore, they are occupied with feeding their wives and children and can simply not afford to hone their family arts, which makes them underachievers within their own profession.

IX

Generally speaking, in days gone by, when we used to receive orders direct from our lords, [the men of Koka] were given large amounts of money in payment for their missions, as much as they required and even if they spent hundreds of Ryo gold coins and if they [the ninja] are asked 'on what was the money spent', they did not need to divulge the answer. This was the normal way for shinobi no mono both in Japan and China and they did not have to balance any accounts. Thus the better shinobi no mono were for missions as spies, then the more money they were supplied. This made everyone do their best to perform great shinobi arts but in modern days, there is simply no demand and they have no opportunity to earn extra money outside of their regular allowances. Therefore, both skilled and lesser skilled [shinobi], are financially poor, thus there are no shinobi no mono that now make constant efforts to become great.

X

As [the arts of the ninja] are a matter of extreme secrets, when teaching those people who are in the same school, the traditions [of the school] are taught so discreetly that they are found in a setting of three or even five stages or levels. In principle, at the first or second stage, you should teach unimportant or useless things to determine whether the person is trustworthy or not, or what abilities he has; if you find him to be serious and resourceful, then you may give him all the traditions. Such a person is only one in a thousand people and the other nine hundred and ninety nine are just taught the first to second stage, which are made to look important and which makes them think they have learned all the traditions of shinobi. However, because of this way, the school is now managed by those who know only the first or second stages of the arts, yet think they have the complete skills, which hinders the progress of our arts; therefore, you must understand that this system is now pointless.

XI

As most students in a school are only at the level as described above, they tend to think that shinobi no jutsu is almost the same as skills of illusion, but there is only one in a multitude who can master the arts to such a deep level that they truly understand that the teachings of the Gokan[5] five types of spy are of the utmost importance and of vital use. All this happens because the teacher does not teach the traditions openly and the students only have limited capabilities.

Though such techniques as hiding yourselves, disguise, crossing over a pond or marsh, crossing a river, climbing a wall or infiltrating a gate are included in the skills of shinobi, they are only trivial and rarely of service to you, making it hard to achieve a great achievement with just these skills alone. Though the Gokan five types of spy are comprised of all that is essential, no one teaches it in detail [any more] and as a result even ambitious people in the school think that such minor skills [as mentioned above] are all secret skills and ways that are important for a shinobi. Because of this, eventually, the skills of shinobi will be become lesser and find limits, this is a certainty.

XII

The difference today is that people are culturally enlightened and tend to learn academically and be more reasonable. Many people do not use or learn hidden secret skills which have been passed down from ancient times, for example; Mitsume and Kikitsume 見詰聞詰 'listening and hearing with intensity', Yojigakure 楊枝カクレ 'Toothpick Hiding', Karamitsume 搦ミツメ 'Catching Finger nail'[6] are regarded as unreasonable and magical and even heretical skills and people often follow up by quoting the following saying: 'There should be nothing mystical about any righteous path'. This retort of theirs is truly an example of the saying, 'Shallow cleverness prevents you from travelling on the righteous path.' In those secret ways or skills invented by ancient people there are numerous things that look unreasonable or unrighteous to the eyes of modern people but in fact have actual benefits and miraculously do work. These miraculous skills seem unreasonable, but this is because others do not reach down deep inside of the way of these things, but you must note, there is reason found within these ancient ways [if you understand the truth of them]. Not having understood such deep reasoning, many people do not believe and abandon those skills thinking they are evil and magical.

XIII

As these traditions are kept so secret, it is often the case that they are discontinued or passed down in a very wrong way because of their very secrecy.

Considering the above thirteen reasons, it can be safely said that there is a huge difference in skill levels found within those who call themselves Koka Shinobi no mono and the main differences can be seen between those who are above 50 years old and under 50 years old, but this has always been the case.[7] Eventually, in the future it will turn out that only one of every ten thousand people can serve well when any emergency arises. Those Daimyo who keep shinobi no mono will only retain them for the name of Koka and for them simply being from Koka. Those who are employed only for that reason are doing nothing more than selling the name of Koka and when such an emergency arises, they will not be able to serve in any way. And, if at such a critical point they cannot perform, it will result not only in his ruin but it could also lead to serious danger for that province. Alongside this, such an event would be a great dishonour for the traditions of Koka and cause the decline of the line itself.

Therefore, those who receive my teachings should be aware of the above reasons before anything and keep training without deviating from the ways or principles of the ancient traditions of this school and should achieve great feats.

Therefore, I here leave these articles recorded to give a warning to those in our school and for the consideration of the future prospects [of Koka ninja].

This document was written on the fifteenth day of the first lunar month in the period of Kyoho 4 [1719]. Also, all the important points from Hara Yuken Yoshifusa's [ninja] scroll called the 'Seikenroku', and the associated oral traditions have been passed on and have all been given, together with a certificate of qualification [to me, Chikamatsu the transcriber of these words].

On this very day mentioned above, my Master dictated this Future Account and I wrote down every word he said, word for word with due respect. On the sixteenth day of the month, I visited my Master to thank him for this and he again told me to record his words which are about a further two matters and are placed here in the following text. This also I wrote down word for word without any difference. These are all essential points of the school and should not be neglected in any way.

Employing or being employed

Those who employ [shinobi] are tacticians or lords while those who are employed are shinobi no mono and even today there are only ten out of one hundred who can truly serve [as ninja]. Alongside this, there might be only one good tactician or lord out of a multitude who is able to utilise his agents very well and this is because good tacticians or lords need to be well versed in military skills and have a grasp of the subtle essence required in employing spies, and should also have a good knowledge of the specific skills of the shinobi. If this is not the case, they cannot judge exactly if a certain action is feasible or not when using shinobi, because of this factor, just under five out of ten [missions] will fail to be successful. [Usually] those [people] who are capable and can determine if a plan is reliable or not do not hold the authority to actually utilise such skills. Thus it can be said that there are only one in a multitude of lords or tacticians who have the understanding required. Also, if they use [shinobi no mono] in a wrong way, it could threaten the survival of a country or cause the destruction of an entire army. But all that a shinobi no mono can do when used in any way is to perform the best he can to fulfil the orders given to him by his superior.

Being a shinobi appears to be difficult but it is in fact not as difficult as it seems, for if his life is endangered, nothing other than his life will be lost. All throughout the history of Japan and China, the lords who used [shinobi] very well were Jinbei, Masamoto, and later on, the late Lord Toshogu [Tokugawa Ieyasu]. These are the only three throughout Japan and China who actually used [ninja] well but there are a countless number of lords who were defeated and killed as they could not use their agents properly. At the battle of Sekigahara, from the beginning and right through to the end, the spies who served Toshogu did very well and fulfilled their aims each and every time, so that many of their secret ways or skills have been passed down in our traditions, while the shinobi who served Lord Ishida Mitsunari [Tokugawa's opponent] used the arts wrongly and could not succeed at every point so they were defeated and ruined. This art [of using shinobi] seems to have been the only point that mattered and was integral to Mitsunari's outcome and was the key factor as to whether the Battle of Sekigaharawas was won or lost.

[As you can see from the above] if the person who uses [shinobi] is not proficient then it is useless and is exactly as the saying, 'A horse who could run one thousand Ri cannot be found without a man who can recognise talent [Hakuraku].' This is exactly the case with a good shinobi, if there is nobody who can use him very well.

Be sure to remember this saying, 'Hawks and musha-warriors will perform dependent upon the person who uses them.'

[*Postscript by Chikamatsu*]
For your information, during the battle of Sekigahara, there were secret episodes which utilised the traditions of Koka and it was prohibited for anyone to write these episodes down, so as to not give them away to others, including the written records about the battle which were spread throughout the country afterwards. Fortunately, my Master has been given the tradition and sometimes he taught these exploits by mouth and from time to time in a series of lectures which ran for three years.

[*Returning to Master Kimura's words*]
The incomparable and indispensable essence found within military ways.
In military tactics the shinobi arts are the incomparable and indispensable to its essence. It might sound as if I am promoting my own interests and should be reproached for self-admiration; however, this is simply not the case and neither is it simply my personal opinion. For proof that these arts are the deepest secrets, you should investigate the thirteenth chapter of *The Art of War* by Sun Tzu. After describing all kinds of military techniques, he discusses the matter of using spies and ends his writing, saying:

> It is the enlightened ruler and the capable general who are able to use the most intelligent ones from within their ranks to be deployed as spies and secret agents so as to achieve the greatest and complete victories in war. Secret operations and espionage activities form an integral part of any military campaign as the planning of strategies and the movement of troops depend heavily upon them.

Sun Tzu emphasises his point by saying 'heavily' or 'complete' as it is of great importance.

As is mentioned in the traditions of 'Niso Daigo' 二相大悟, everything originates from ears and eyes and this is exactly what shinobi do as they travel thousands of Ri across mountains and rivers, serving as ears and eyes and report all they see and hear [to the lord]. Whoever the great generals of all ages were, how could they have gained this intelligence without sending out shinobi? Shinobi are the ears and eyes of the entire army and so 'the movement of troops depend heavily upon them'. Therefore, it can be said this is the deepest essence of warfare, more important than any other arts.

People from most military schools in Japan are barely literate and cannot read *The Art of War* in very fine detail, therefore, they are not aware of what it means and do not think that using spies is essential or important, they even think it is no better than shinobi-stealing[8] 竊盜, which is actually in complete contradiction to what Sun Tsu teaches in his work, and for this reason you should try to understand fully the subtle meaning of what

Sun Tzu's work means and build the basis for the success of your entire army from that chapter. Those who do not understand the true and deepest of secrets of military warfare should know that they can all be found in Sun Tzu's *The Art of War* and those who do not know this are not worth conversing with on the topic of warfare.

I write here for your information, that the lectures on the chapter of using spies in *The Art of War* has been complied into a series of ten lectures and these were recorded with commentary.

This future account and the above two important matters were both dictated and checked by my Master, and copied on the third day of second lunar month and in the year of the Kinoto-Boar (1719) in Bishu (Owari province) and transcribed by me, Nogen (Chikamatsu) of the Renpeido soldier training complex.

Further transcribed in the year of Kinoto-Ox in the eighth month of 1805 in Osaka by Suzuki Sadayoshi, a warrior of Bishu and given to Mizuno Gentadamichi.

NOTES

1 Throughout the entire text, the author talks of the two principles of constant communication and the idea of keeping the *shinobi* teachings within only those from Koka.
2 The author constantly switches from present to past tense and the translation attempts to follow his lead.
3 Sengoku Period.
4 There is an ambiguous idiom here, however by context it means unused.
5 The author is saying that many people get caught up in the shallow ends of deception and that the truth shall be found in the way of the five spies, the concept initiated by Sun Tzu.
6 Here the 'catching' has a connotation of a winding motion.
7 He does not seem to be referencing the end of the Sengoku Period generation but the difference in understanding between the old and young. Fujibayashi also states that older shinobi are better than younger ones.
8 This sentence should not be missed by any student of *ninjutsu*. Kimura is inferring that those who perform this form of shinobi are only thieves. The difference is in the use of the ideogram, the one here is found in various writings and sometimes together with the classical ideogram for shinobi. This implies that there are two types of shinobi action.

APPENDICES

Appendix A

The Eboshi-ori Play

Article twenty-four of the first scroll in this book, Taimatsu no Urate no Daiji, *The Principle of Divination by Torches* discusses a play of the Japanese theatre form known as Noh. The article states that Divination by Torches and the concept of 'wind' as idiomatic expressions used by shinobi both appear in it. The story follows the young and soon to be famous warrior Yoshitsune and his struggle against the infamous thief Kumasaka Chohan. To aid in understanding the article from the text we have translated and precised the relevant section concerning Divination by Torches from the play.

Yoshitsune[273] was a son of the noble samurai Minamoto family. Yoshitsune was left under the care of Taira no Kiyomori at Kurama temple when he was very young, the Taira clan were the enemy of the Minamoto family and therefore he grew up among the enemy. When he became 16 or 17 years old Yoshitsune escaped the temple and fled to Mutsu province and on his way he met rich merchant brothers and asked if he could travel together with them. On their way they came to Akasaka in Mino province, and stayed at an inn. The master of the inn informed them that a group of bandits had heard that the rich merchants were staying at the inn and were therefore planning a raid that very night. The merchants were fearful but Yoshitsune assured them that he would get rid of the bandits for them.

Yoshitsune waited for them in the house with the front door open. The three subordinates of Kumasaka Chohan[274] came with torches in hand. They were told by Kumasaka Chohan to scout out and discover the situation in the inn. The first one crossed over the outside wall and entered the grounds but while investigating he found Yoshitsune standing before him. Startled, he ran back to his colleagues.

Next the three [scouts] wanted to find out just what or who he saw and decided to throw a torch into the ground around that area. The second of them went in through the gate with a torch and found Yoshitsune, upon which he threw the torch at him, but Yoshitsune stepped upon on the torch and doused the flames without difficulty, and thus [the scout] ran back to his allies.

Next the third went inside with a torch and threw it at Yoshitsune but Yoshitsune cut it down out of the air with his Tachi longsword.

Finally, the original scout moved in with a torch but Yoshitsune took the torch from him and cut him upon his shoulder and threw the torch back at them. Upon this the three ran away, back to the main body of their group.

Kumasaka Chohan sent a force of seventy people but many of them were killed or injured and so they retreated. Next, Kumasaka Chohan asked a captain what had happened and considered who it was that was defending the lodgings in such a way.

The conversation went thus:

Kumasaka Chohan: 'Why were those men defeated and was there a strong force inside?'

Captain: 'Yes sir, the men said that the force inside was so strong and that our allies were killed or injured.'

Kumasaka Chohan: 'That is strange. There should be no one other than just the [rich] merchants. What was this man like?'

Captain: 'By the light of a torch, they saw a 16- or 17-year-old boy fighting with a Kodachi sword, he was moving as lightly as a butterfly or a bird.'

Kumasaka Chohan: 'What happened with [our scouts], Suribachi Taro and his brothers?'

Captain: 'They went there first in the role of the captain of Hiburi [Hiburi no Oyakata] fire squad but the boy fought with them and beheaded one of them with only one cut.'

Kumasaka Chohan: 'What is this, what do you say! What happened? The [three] scouts are worth more than fifty or even one hundred people. But he cut them down, who in hell is this boy I wonder?'

Captain: 'Takase no Shiro – who was one of the bandits – said that he thought that luck was not our side tonight and withdrew all seventy of his men.'

Kumasaka Chohan: 'Takase is always a coward. But tell me, what of the divination by torches?'

Captain: 'Well, the first torch was extinguished when it was stepped upon. The second torch was cut down out of the air and the third torch was thrown back and all three were extinguished.'

Kumasaka Chohan: 'That is a serious matter. Divination by torches originally means: the god of war for the first torch, the luck [of the present time] for the second torch and our life by the third torch and now as I hear that all three were put out, I am concerned for the night raid we have planned.'

Captain: 'As you say, it seems it is difficult even for a demon to deal with this situation. I think we had better retreat.'

Kumasaka Chohan: 'True, it is not worth risking our lives over, let us retreat.'

Captain: 'Agreed.'

Kumasaka Chohan (after contemplation): 'No, I now think we should not retreat after all. Am I not Kumasaka Chohan, the one who is well known to everyone all along Tokaido highway? If I fail to carry out this night raid then it will be beyond embarrassment, so go my men, go and attack!'

At once they rushed through the gate and Yoshitsune did battle with them on his own and won, [as that night] many bandits fought among themselves. Yoshitsune succeeded in killing Kumasaka Chohan in the end.

The use of the shield

Appendix B

THE COLLECTED SKILLS OF IGA AND KOKA

The following two lists are a compilation of all the skills in the above manuals; they have been produced in alphabetical order in both English and Japanese.

ENGLISH LIST

The skills of Iga and Koka in alphabetical order according to their English translation:

1 Concerning the Days to be Avoided – En'nichi no Koto 厭日之事
2 Concerning In and Yo – In Yo Tomoni Majiwaru Koto 陰陽俱錯事
3 Concerning Ten'ichijin – Ten'ichijin no Koto 天一神之事
4 Concerning Tengushin – Tengushin no Koto 天宮神之事
5 Concerning the Days of Doko – Dokonichi no Koto 道虚日之事
6 Concerning the Days When You Should Shoot Arrows [to Dispel Evil] – Yahanachibi no Koto 箭放日事
7 Concerning the Direction of Sashigami - Sashigami no Kata no Koto 指神方之事
8 Concerning the Seasons and the Changing Power of Five Phases – Shiki Oso no Koto 四季王相之事
9 Concerning the Star of Hagun – Hagunsei no Koto 破軍星之事
10 Concerning the Three Mansions – Mishuku no Koto 三宿之事
11 Concerning Unlucky Days for Travelling – Toku e Iku wo Imu Koto 忌遠行事
12 Concerning Unlucky Days for Travelling at Night – Yoru Iku wo Imu Koto 忌夜行事
13 Guiding Principles of Behaviour for Shinobi – Shinobi Mimochi no Koto 忍身持之事
14 Hints about the Clothing for a Shinobi – Shinobi Kokoromochi no Koto 忍心持之事
15 Hints for Before and After [the Festivals of] Jizo and Yakushi – Jizo Yakushi no Mae Ushiro no Kokoroe 地蔵薬師ノ前ウシロノ心得
16 How to Protect Yourself from the Cold – Kogoezaru no Daiji 不凍大事
17 Principles Concerning the Bedroom – Neya no Daiji 寝屋ノ大事
18 Principles on Fire Handlers – Hiburi Hitoboshi no Daiji 火フリ火トホシノ大事
19 Secret Magic Arts Used By Shinobi When Venturing Out – Shinobi Ikubeki Hiho no Koto 忍可行秘法之事
20 Teachings Concerning the Traditions on Introversive Chi and Outgoing Chi – Uchinoki Sotonoki Kunden 内之氣外之氣訓傳
21 The Art Concerning Crossing by Animals and the Luck or Bad Luck it Brings - Shisoku Au Kikkyo no Koto 四足逢吉凶之事
22 The Art of Avoiding Being Hated by Dogs of the Corners – Kado no Inu nimo Nikumarenu Koto 角ノ犬ニモ悪マレヌ事
23 The Art of Before, Middle and After – Zen Chu Go no Koto 前中後之事

24 The Art of Blinding Power – Seimyosan no Koto 井妙散之事
25 The Art of Buddha Hiding – Hotoke Gakure no Koto 佛カクレノ事
26 The Art of Bush Hiding – Shiba Gakure no Koto 柴隠レノ事
27 The Art of Carrying Your Tools – Dogu Mochiyo no Koto 道具持様之事
28 The Art of Catching your Breath – Ikiai no Koto 息合之事
29 The Art of Changing Direction – Hogaku Henka no Koto 方角変化之事
30 The Art of Choosing [people] – Eramu Koto 撰ム事
31 The Art of Choosing a Positive Hour – Tokidori no Koto 時取之事
32 The Art of Confusing Your Tracks – Midareashi Gutsu no Koto 乱足沓之事
33 The Art of Creation and Destruction Cycles – Sosho Sokoku [no] Koto 相性相尅事
34 The Art of Crossing a River in the Daytime or at Night Time – Chuya Kawagoe no Koto 昼夜川越之事
35 The Art of Destroying an Army – Ninju Kuzushi no Koto 人数崩之事
36 The Art of Destruction and Creation Cycles – Sokoku Sosho [no] Koto相尅相生事
37 The Art of Discovering Fire – Jikenhi no Koto 自見火之事
38 The Art of Dog Hiding – Inu Gakure no Koto 犬カクレノ事
39 The Art of Dreaming Powder – Ichimusan no Koto 一夢散之事
40 The Art of Hiding Your Name to Aid a Great Victory – Taisho no Tame ni Na wo Oshimu Koto 大勝ノ為ニ名ヲ惜ム事
41 The Art of Hunger Pills – Hyorogan no Koto 兵粮丸之事
42 The Art of Identifying Suspicious Woods – Shinrin Fushin no Koto 森林不審之事
43 The Art of In Shinobi – In Shinobi no Koto 陰忍之事
44 The Art of In War Curtains – In Maku no Koto 陰幕之事
45 The Art of Incense Smoke – In Connection With Luck – Koen Zen'aku no Koto 香煙善悪之事
46 The Art of Infiltrating a Vast Army – Taigun no Naka wo Toru Koto 大軍中通事
47 The Art of Internal Wind and External Wind – Uchi no Kaze Soto no Kaze to Iu Koto 内ノ風外ノ風ト云事
48 The Art of Judging How Many People You Should Use – Daisho no Kokoroe Arubeki Koto 大小ノ心得アルヘキ事
49 The Art of Judging How Much Money Should be Used – Tasho no Kokoroe Arubeki Koto 多少ノ心得アルヘキ事
50 The Art of Knowing How the Enemy Have Prepared Their Castle – Tekijo Yoi Shiru Koto 敵城用意知事
51 The Art of Knowing if a Situation is Lucky or Unlucky When Venturing out on a Shinobi Mission – Shiniobi Ideru Kikkyo wo Shiru Koto 忍出吉凶知事
52 The Art of Knowing Where to Hide When You Encounter the Enemy – Teki ni Ai Kakuredokoro no Koto 敵逢隠所事
53 The Art of Knowing Which Lunar Mansion by the First Day of the Month – Hi no Obun no Koto 日王分之事
54 The Art of Knowing Which Lunar Mansion is Most Important by the Hour – Toki no Ojun no Koto 時王順之事
55 The Art of Knowing Which Lunar Mansion Is Prominent In Which Year – Toshi no Oi no Koto 年王位之事
56 The Art of Leaf Hiding – Konoha Gakure no Koto 木ノ葉カクレノ事
57 The Art of Leaving and Entering – Hi no Deiri no Koto 日出入之事

58 The Art of Leaving Your Name for the Time of Victory – Godo no Kachi ni Na wo Nokosu Koto 後途ノ勝ニ名ヲ残ス事
59 The Art of Luck in Relation to Crossing by a Dead Body – Shibito Au Kikkyo no Koto 死人逢吉凶之事
60 The Art of Luck In Relation to Various Birds – Shocho Kikkyo no Koto 諸鳥吉凶之事
61 The Art of Luring Somebody Out – Yobidashi no Jutsu 呼出シノ術
62 The Art of Making a Conversation Blossom – Ji ni Hana wo Sakaseru Ben 辞ニ花ヲ咲セル辨
63 The Art of Making Handheld Shields – Tedate Koshiraeyo no Koto 手楯拵様之事
64 The Art of Marsh Floating Shoes – Numa Ukigutsu no Koto 沼浮沓之事
65 The Art of Night Raids – Youchi no Koto 夜討之事
66 The Art of Not Casting Stones on a Pitch Black Night – Yami no Yo ni Tsubute Utsubekarazaru Koto 闇ノ夜ニ礫ウツヘカラサル事
67 The Art of Not Shooting an Arrow if you are Going to Miss – Atarazaru Ya ha Hanatazu no Koto 不中矢ハ不放事
68 The Art of Passing Dead Bodies – Shibito Tooriyo no Koto 死人通様之事
69 The Art of Paying Attention to Various Insects – Shochu Kokorozuke no Koto 諸虫心付之事
70 The Art of Questioning – Kaeri toi no Koto カヘリ問ノ事
71 The Art of Respecting the Gods – Shinki no Koto 神貴之事
72 The Art of Shinobi Clothing – Shinobi Irui no Koto 忍衣類之事
73 The Art of Ship Destruction – Funakuzushi no Koto 舟崩之事
74 The Art of Taking the Pillow – Makura wo Toru Koto 枕ヲトル事
75 The Art of the Basket Fire – Kagohi no Koto 籠火之事
76 The Art of the Entering Death Fire – Irishibi no Koto 入死火之事
77 The Art of the Fast-shelter – Hayagoya no Koto 早小屋之事
78 The Art of the Gods, of Rivers and of Floods – Kahaku Omizu Gami no Koto 河伯大水神之事
79 The Art of the Hand Throwing Fire – Tebiya no Koto 手火矢之事
80 The Art of the Moon on the Water – Suigetsu no Daiji 水月之大事
81 The Art of the Relationship between a Given Element in Association with the Twelve Signs of the Chinese Zodiac and the Ten Celestial Stems – Shikan Oshu no Koto 支干王主之事
82 The Art of the Sliced Torch – Kiri Taimatsu no Koto 切松明之事
83 The Art of the Smoke Signal – Noroshi no Koto 狼煙之事
84 The Art of the Strike Fire Signal – Uchinoroshi no Koto 打燧之事
85 The Art of the Three Fold Fire – Mienohi no Koto 三重火之事
86 The Art of the Tied Bridge – Yuibashi no Koto 結橋之事
87 The Art of the Toothpick Fire – Yojihi no Koto 楊枝火之事
88 The Art of the Tsutsunohi Cylinder Fire – Tsutsunoni no Koto 筒火之事
89 The Art of the Universal Medicine – Kimyosan no Koto 貴妙散之事
90 The Art of the Water-Proof Torch – Mizutaimatsu no Koto 水松續之事
91 The Art of Turret Collapsing – Yagura Otoshi no Koto 櫓落之事
92 The Art of Tying and Distributing Caltrops – Hishi Musubi Kubariyo no Koto 菱結配様之事

93 The Art of Unlucky Days for Returning From a Mission – Kimonichi no Koto 帰亡日之事
94 The Art of Unlucky Days for Venturing Out on a Mission – Omonichi no Koto 往亡日之事
95 The Art of Using Dust and Fine Grains – Mijin Tsukaiyo no Koto 微塵遣様之事
96 The Art of Using Powder to Keep Sleep at Bay – Miyosan no Koto 未用散之事
97 The Art of Using Rokuzen Spikes – Rokuzen Tsukaiyo no Koto 六膳遣様之事
98 The Art of Using the Kururukagi Key – Kururukagi Tsukaiyo no Koto 枢鎰遣様之事
99 The Art of Using the Roll-away Ladder – Makibashi Tsukaiyo no Koto 巻橋遣様之事
100 The Art of Yo Shinobi – Yo Shinobi no Koto 陽忍之事
101 The Art of Yo War Curtains – Yo Maku no Koto 陽幕之事
102 The Arts of Mitsume and Kikitsume – Mitsume Kikitsume no Daiji 見詰聞詰之大事
103 The Most Important Lunar Mansion Connected to Each Month - Tsuki no Oza no Koto 月王座之事
104 The Opposite to the Above Unlucky Days – Entainichi 厭對日
105 The Origin of Shinobi – Shinobi no Kigen 忍之起原
106 The Principle [of Retaining] an In-body – Mi wo In no Daiji 身陰之大事
107 The Principle of Connecting with the Hearts [of Other Shinobi] – Kokoro Yui no Daiji 心結之大事
108 The Principle of Crossing Water – Suichu no Daiji 水中之大事
109 The Principle of Divination by Torches – Taimatsu no Urate no Daiji 松明之占手之大事
110 The Principle of Double Sowing – Shikimaki no Daiji 重播之大事
111 The Principle of In & Yo Shinobi – In Yo Shinobi no Daiji 陰陽忍之大事
112 The Principle of Leaf Hiding – Konoha Gakure no Daiji 木之葉隠之大事
113 The Principle of Mist – Kasumi no Daiji 霞之大事
114 The Principle of Secret Connection – Ura-awase no Daiji 占合之大事
115 The Principle of Shinobi Identification – Shinobi Araware no Daiji 忍顕大事
116 The Principle of Sleeping Quarters – Neya no Daiji 寝屋之大事
117 The Principle of the Dream Pillow – Yumemakura no Daiji 夢枕之大事
118 The Principle of the Grass Bundle – Kusazuto no Daiji 草苞之大事
119 The Principle of the Greater-Shinobi – Ooshinobi no Daiji 大忍之大事
120 The Principle of the Mudra of Mist – Kiri no In no Daiji 霧之印ノ大事
121 The Principle of the Pivot – Kaname no Daiji 要之大事
122 The Principle of the Seven Emotions – Shichiji no Daiji 七字之大事
123 The Principle of the Three-way Connection – Miawase no Daiji 三合ノ大事
124 The Principle of Throwing Stones – Soratsubute no Daiji 空礫之大事275
125 The Principle of Understanding of the Character of People – Chi Jo no Daiji 知情之大事
126 The Secret art of Correct Connection - Seigo no Hijutsu 正合之秘術
127 The Seven ri torch - Shichiri Taimatsu no Koto 七里炬之事
128 The Shinobi Art of Knowing – RoadsShinobi Michimi no Koto 忍路見之事
129 The Shinobi Art of Walking – Shinobi Michi Fumiyo no Koto 忍道踏様之事
130 The Skill of Hiding Your Identity - Katachi wo Kakusu Jutsu 形ヲ隠ス術
131 The Skill of Interrogation by Sleep Deprivation – Utsutsudoi no Jutsu 現問ノ術
132 The Skill of Poison Which Will In Turn Become Medicine – Dokuyaku Hanshite Kusuri tonaru Jutsu 毒薬反シテ薬トナル術

133 The Skill of Sending an Important Message – Daiji no Tsukai no Jutsu 大事ノ使ノ術
134 The Skill of the Arm-restraining Trap – Ude Karami no Jutsu 腕カラミノ術
135 The Skill of the Foothold Trap – Ashi Karami no Jutsu 足カラミノ術
136 The Tradition of Cattles and Horses – Gyuba no Tsutae 牛馬ノ傳
137 The Tradition of Echoes – Yamabiko no Tsutae 山彦之傳
138 The Tradition of Fleas and Lice – Nomi Shirami no Tsutae 蚤虱之傳
139 The Tradition of Flying Birds – Hicho no Tsutae 飛鳥之傳
140 The Tradition of Great Stability – Dai Suwari no Tsutae 大スワリノ傳
141 The Tradition of Hiding from Sight – Magakure no Tsutae マカクレノ傳
142 The Tradition of No Sword – Muto Den 無刀傳
143 The Tradition of Riding on the Wind – Kaze ni Noru no Tsutae 風ニ乗ルノ傳
144 The Tradition of Separation – Ichiri no Tsutae 一離之傳
145 The Tradition of the Letter in a Fire – Kachujo no Tsutae 火中状ノ傳
146 The Tradition of the Movement of Clouds – Unko no Tsutae 雲行之傳
147 The Tradition of the Rear Attack – Wakekiri no Tsutae ワケキリノ傳
148 The Traditions of Attacking the Front – Omote Uchi no Tsutae 面討之傳
149 The Traditions of Exhausting the Enemy – Tsukarakashi no Tsutae ツカラカシノ傳
150 The Traditions of Four Ways of Knowing – Shichi no Tsutae 四知之傳
151 The Traditions of Purse Web Spider – Anagumo Jigumo no Tsutae 穴蜘蛛地蜘蛛ノ傳
152 The Traditions of Small Castles - Kojiro no Tsutae 小城之傳
153 The Traditions of the Fox and the Wolf - Koro no Tsutae 狐狼ノ傳
154 The Traditions of the Meaning of the Ideogram Shinobi 忍 – Shinobi no Kunden 忍之訓傳
155 The Traditions of the Mosquito and the Fly - Ka Hae no Tsutae 蚊蠅之傳
156 The Traditions of the Spider - Kumo no Tsutae 蜘蛛之傳
157 The Traditions of using Voice as Guide – Koe wo Shirube no Tsutae 声ヲ知ルヘノ傳
158 The Traditions of Walking Main Highways – Michi wo Michi to Subeki Tsutae 道ヲ道トスヘキ傳
159 The Traditions of Writing a Series of Fake Letters – Chirashi Gaki no Tsutae チラシ書ノ傳
160 The Traditions on Introversive Chi and Outgoing Chi – Uchinoki Sotonoki no Tsutae 内之氣外之氣之傳
161 The Trick of Cut Fuse - Kiri Hinawa no Shosa 切火縄ノ所作

JAPANESE LIST

The skills of Iga and Koka in alphabetical according to their Japanese pronunciation:

1 Anagumo Jigumo no Tsutae 穴蜘蛛地蜘蛛ノ傳 The Traditions of Purse Web Spider
2 Ashi Karami no Jutsu 足カラミノ術 The Skill of the Foothold Trap
3 Atarazaru Ya ha Hanatazu no Koto 不中矢ハ不放事 The Art of Not Shooting an Arrow if you are Going to Miss
4 Chi Jo no Daiji 知情之大事 The Principle of Understanding the Character of People

5 Chirashi Gaki no Tsutae チラシ書ノ傳 The Traditions of Writing a Series of Fake Letters
6 Chuya Kawagoe no Koto 昼夜川越之事 The Art of Crossing a River in the Daytime or at Night Time
7 Dai Suwari no Tsutae 大スワリノ傳 The Tradition of Great Stability
8 Daiji no Tsukai no Jutsu 大事ノ使ノ術 The Skill of Sending an Important Message
9 Daisho no Kokoroe Arubeki Koto 大小ノ心得アルヘキ事 The Art of Judging How Many People You Should Use
10 Dogu Mochiyo no Koto 道具持様之事 The Art of Carrying Your Tools
11 Dokonichi no Koto 道虚日之事 Concerning the Days of Doko
12 Dokuyaku Hanshite Kusuri tonaru Jutsu 毒薬反シテ薬トナル術 The Skill of Poison Which Will In Turn Become Medicine
13 En'nichi no Koto 厭日之事 Concerning Days to be Avoided
14 Entainichi 厭對日 The Opposite to the Above Unlucky Days
15 Eramu Koto 撰ム事 The Art of Choosing [people]
16 Funakuzushi no Koto 舟崩之事 The Art of Ship Destruction
17 Godo no Kachi ni Na wo Nokosu Koto ノ勝ニ名ヲ残ス事 The Art of Leaving Your Name for the Time of Victory
18 Gyuba no Tsutae 牛馬ノ傳 The Tradition of Cattles and Horses
19 Hagunsei no Koto 破軍星之事 Concerning the Star of Hagun
20 Hayagoya no Koto 早小屋之事 The Art of the Fast-shelter
21 Hi no Deiri no Koto 日出入之事 The Art of Leaving and Entering
22 Hi no Obun no Koto 日王分之事 The Art of Knowing Which Lunar Mansion by the First Day of the Month
23 Hiburi Hitoboshi no Daiji 火フリ火トホシノ大事 Principles about Fire Handlers
24 Hicho no Tsutae 飛鳥之傳 The Tradition of Flying Birds
25 Hishi Musubi Kubariyo no Koto 菱結配様之事 The Art of Tying and Distributing Caltrops
26 Hogaku Henka no Koto 方角変化之事 The Art of Changing Direction
27 Hotoke Gakure no Koto 佛カクレノ事 The Art of Buddha Hiding
28 Hyorogan no Koto 兵粮丸之事 The Art of Hunger pills
28 Ichimusan no Koto 一夢散之事 The Art of Dreaming Powder
30 Ichiri no Tsutae 一離之傳 The Tradition of Separation
31 Ikiai no Koto 息合之事 The Art of Catching your Breath
32 In Maku no Koto 陰幕之事 The Art of In War Curtains
33 In Shinobi no Koto 陰忍之事 The Art of In Shinobi
34 In Yo Shinobi no Daiji 陰陽忍之大事 The Principle of In & Yo Shinobi
35 In Yo Tomoni Majiwaru Koto 陰陽倶錯事 Concerning In and Yo
36 Inu Gakure no Koto 犬カクレノ事 The Art of Dog Hiding
37 Irishibi no Koto 入死火之事 The Art of the Entering Death Fire
38 Ji ni Hana wo Sakaseru Ben 辞ニ花ヲ咲セル辨 The Art of Making a Conversation Blossom
39 Jikenhi no Koto 自見火之事 The Art of Discovering Fire
40 Jizo Yakushi no Mae Ushiro no Kokoroe 地蔵薬師ノ前ウシロノ心得 Hints for Before and After [the Festivals of] Jizo and Yakushi
41 Ka Hae no Tsutae 蚊蝿之傳 The Traditions of the Mosquito and the Fly

42 Kachujo no Tsutae 火中状ノ傳 The Tradition of the Letter in a Fire
43 Kado no Inu nimo Nikumarenu Koto 角ノ犬ニモ悪マレヌ事 The Art of Avoiding Being Hated by Dogs of the Corners
44 Kaeri toi no Koto カヘリ問ノ事 The Art of Questioning
45 Kagohi no Koto 籠火之事 The Art of the Basket Fire
46 Kahaku Omizu Gami no Koto 河伯大水神之事 The Art of the Gods, of Rivers and of Floods
47 Kaname no Daiji 要之大事 The Principle of the Pivot
48 Kasumi no Daiji 霞之大事 The Principle of Mist
49 Katachi wo Kakusu Jutsu 形ヲ隠ス術 The Skill of Hiding Your Identity
50 Kaze ni Noru no Tsutae 風ニ乗ルノ傳 The Tradition of Riding on the Wind
51 Kimonichi no Koto 帰亡日之事 The Art of Unlucky Days for Returning From a Mission
52 Kimyosan no Koto 貴妙散之事 The Art of the Universal Medicine
53 Kiri Hinawa no Shosa 切火縄ノ所作 The Trick of the Cut Fuse
54 Kiri no In no Daiji 霧之印ノ大事 The Principle of the Murda of Mist
55 Kiri Taimatsu no Koto 切松明之事 The Art of the Sliced Torch
56 Koe wo Shirube no Tsutae 声ヲ知ルヘノ傳 The Traditions of using Voice as Guide
57 Koen Zen'aku no Koto 香煙善悪之事 The Art of Incense Smoke – In Connection With Luck
58 Kogoezaru no Daiji 不凍大事 How to Protect Yourself from the Cold
59 Kojiro no Tsutae 小城之傳 The Traditions of Small Castles
60 Kokoro Yui no Daiji 心結之大事 The Principle of Connecting with the Hearts [of Other Shinobi]
61 Konoha Gakure no Daiji 木之葉隠之大事 The Principle of Leaf Hiding
62 Konoha Gakure no Koto 木ノ葉カクレノ事 The Art of Leaf Hiding
63 Koro no Tsutae 狐狼ノ傳 The Traditions of the Fox and the Wolf
64 Kumo no Tsutae 蜘蛛之傳 The Traditions of the Spider
65 Kururukagi Tsukaiyo no Koto 枢鎰遣様之事 The Art of Using the Kururukagi Key
66 Kusazuto no Daiji 草苞之大事 The Principle of the Grass Bundle
67 Magakure no Tsutae マカクレノ傳 The Tradition of Hiding from Sight
68 Makibashi Tsukaiyo no Koto 巻橋遣様之事 The Art of Using the Roll-away Ladder
69 Makura wo Toru Koto 枕ヲトル事 The Art of Taking the Pillow
70 Mi wo In no Daiji 身陰之大事 The Principle [of Retaining] an In-body
71 Miawase no Daiji 三合ノ大事 The Art of the Three-way Connection
72 Michi wo Michi to Subeki Tsutae 道ヲ道トスヘキ傳 The Traditions of Walking Main Highways
73 Midareashi Gutsu no Koto 乱足沓之事 The Art of Confusing Your Tracks
74 Mienohi no Koto 三重火之事 The Art of the Three Fold Fire
75 Mijin Tsukaiyo no Koto 微塵遣様之事 The Art of Using Dust and Fine Grains
76 Mishuku no Koto 三宿之事 Concerning Three Mansions
77 Mitsume Kikitsume no Daiji 見詰聞詰之大事 The Principles of Mitsume and Kikitsume
78 Miyosan no Koto 未用散之事 The Art of Using Powder to Keep Sleep at Bay
79 Mizutaimatsu no Koto 水松續之事 The Art of the Water-Proof Torch
80 Muto Den 無刀傳 The Tradition of No Sword

81 Neya no Daiji 寝屋ノ大事 Important Points Concerning the Bedroom
82 Neya no Daiji 寝屋之大事 The Principle of Sleeping Quarters
83 Ninju Kuzushi no Koto 人数崩之事 The Art of Destroying an Army
84 Nomi Shirami no Tsutae 蚤虱之傳 The Tradition of Fleas and Lice
85 Noroshi no Koto 狼煙之事 The Art of the Smoke Signal
86 Numa Ukigutsu no Koto 沼浮沓之事 The Art of Marsh Floating Shoes
87 Omonichi no Koto 往亡日之事 The Art of Unlucky Days for Venturing Out on a Mission
88 Omote Uchi no Tsutae 面討之傳 The Traditions of Attacking the Front
89 Ooshinobi no Daiji -大忍之大事 The Principle of the Greater-Shinobi
90 Rokuzen Tsukaiyo no Koto 六膳遣様之事 The Art of Using Rokuzen Spikes
91 Sashigami no Kata no Koto 指神方之事 Concerning the Direction of Sashigami
92 Seigo no Hijutsu 正合之秘術 The Secret art of Correct Connection
93 Seimyosan no Koto 井妙散之事 The Art of Blinding Power
94 Shiba Gakure no Koto 柴隠レノ事 The Art of Bush Hiding
95 Shibito Au Kikkyo no Koto 死人逢吉凶之事 The Art of Luck in Relation to Crossing by a Dead Body
96 Shibito Tooriyo no Koto 死人通様之事 The Art of Passing Dead Bodies
97 Shichi no Tsutae 四知之傳 Traditions of Four Ways of Knowing
98 Shichiji no Daiji 七字之大事 The Principle of the Seven Emotions
99 Shichiri Taimatsu no Koto 七里炬之事 The Seven ri torch
100 Shikan Oshu no Koto 支干王主之事 The Art of the Relationship between a Given Element in Association with the Twelve Signs of the Chinese Zodiac and the Ten Celestial Stems
101 Shiki Oso no Koto 四季王相之事 Concerning the Seasons and the Changing Power of Five Phases
102 Shikimaki no Daiji 重播之大事 The Principle of Double Sowing
103 Shinki no Koto 神貴之事 The Art of Respecting the Gods
104 Shinobi Araware no Daiji 忍顕大事 The Principle Shinobi Identification
105 Shinobi Ideru Kikkyo wo Shiru Koto 忍出吉凶知事 The Art of Knowing if a Situation is Lucky or Unlucky When Venturing out on a Shinobi Mission.
106 Shinobi Ikubeki Hiho no Koto 忍可行秘法之事 Secret Magic Arts Used By Shinobi When Venturing Out
107 Shinobi Irui no Koto 忍衣類之事 The Art of Shinobi Clothing
108 Shinobi Kokoromochi no Koto 忍心持之事 Hints about the Clothing for a Shinobi
109 Shinobi Michi Fumiyo no Koto 忍道踏様之事 The Shinobi Art of Walking
110 Shinobi Michimi no Koto 忍路見之事 The Shinobi Art of Knowing Roads
111 Shinobi Mimochi no Koto 忍身持之事 Guiding Principles of Behaviour for Shinobi
112 Shinobi no Kigen 忍之起原 The Origin of Shinobi
113 Shinobi no Kunden 忍之訓傳 The Traditions of the Meaning of the Ideogram Shinobi 忍
114 Shinrin Fushin no Koto 森林不審之事 The Art of Identifying Suspicious Woods
115 Shisoku Au Kikkyo no Koto 四足逢吉凶之事 The Art Concerning Crossing by Animals and the Luck or Bad Luck it Brings
116 Shocho Kikkyo no Koto 諸鳥吉凶之事 The Art of Luck In Relation to Various Birds

117 Shochu Kokorozuke no Koto 諸虫心付之事 The Art of Paying Attention to Various Insects
118 Sokoku Sosho [no] Koto 相尅相生事 The Art of Destruction and Creation Cycles
119 Soratsubute no Daiji 空礫之大事 The Principle of Throwing Stones
120 Sosho Sokoku [no] Koto 相性相尅事 The Art of Creation and Destruction Cycles
121 Suichu no Daiji 水中之大事 The Principle of Crossing Water
122 Suigetsu no Daiji 水月之大事 The Principle of the Moon on the Water
123 Taigun no Naka wo Toru Koto 大軍中通事 The Art of Infiltrating a Vast Army
124 Taimatsu no Urate no Daiji 松明之占手之大事 The Principle of Divination by Torches
125 Taisho no Tame ni Na wo Oshimu Koto 大勝ノ為ニ名ヲ惜ム事 The Art of Hiding Your Name to Aid a Great Victory
126 Tasho no Kokoroe Arubeki Koto 多少ノ心得アルヘキ事 The Art of Judging How Much Money Should be Used
127 Tebiya no Koto 手火矢之事 The Art of the Hand Throwing Fire
128 Tedate Koshiraeyo no Koto 手楯拵様之事 The Art of Making Handheld Shields
129 Teki ni Ai Kakuredokoro no Koto 敵逢陰所事 The Art of Knowing Where to Hide When You Encounter the Enemy
130 Tekijo Yoi Shiru Koto 敵城用意知事 The Art of Knowing How the Enemy Have Prepared Their Castle
131 Ten'ichijin no Koto 天一神之事 Concerning Ten'ichijin
132 Tengushin no Koto 天宮神之事 Concerning Tengushin
133 Toki no Ojun no Koto 時王順之事 The Art of Knowing Which Lunar Mansion is Most Important by the Hour
134 Tokidori no Koto 時取之事 The Art of Choosing a Positive Hour
135 Toku e Iku wo Imu Koto 忌遠行事 Concerning Unlucky Days for Travelling
136 Toshi no Oi no Koto 年王位之事 The Art of Knowing Which Lunar Mansion Is Prominent In Which Year
137 Tsukarakashi no Tsutae ツカラカシノ傳 The Traditions of Exhausting the Enemy
138 Tsuki no Oza no Koto 月王座之事 The Most Important Lunar Mansion Connected to Each Month
139 Tsutsunohi no Koto 筒火之事 The Art of the Tsutsunohi Cylinder Fire
140 Uchi no Kaze Soto no Kaze to Iu Koto 内ノ風外ノ風ト云事 Concerning Internal Wind and External Wind
141 Uchinoki Sotonoki Kunden 内之氣外之氣訓傳 Teachings Concerning the Traditions on Introversive Chi and Outgoing Chi
142 Uchinoki Sotonoki no Tsutae 内之氣外之氣之傳 The Traditions on Introversive Chi and Outgoing Chi
143 Uchinoroshi no Koto 打燧之事 The Art of the Strike Fire Signal
144 Ude Karami no Jutsu 腕カラミノ術 The Skill of the Arm-restraining Trap
145 Unko no Tsutae 雲行之傳 The Tradition of the Movement of Clouds
146 Ura-awase no Daiji 占合之大事 The Principle of Secret Connection
147 Utsutsudoi no Jutsu 現問ノ術 The Skill of Interrogation by Sleep Deprivation
148 Wakekiri no Tsutae ワケキリノ傳 The Tradition of the Rear Attack
149 Yagura Otoshi no Koto 櫓落之事 The Art of Turret Collapsing

150 Yahanachibi no Koto 箭放日事 Concerning the Days When You Should Shoot Arrows [to Dispel Evil]
151 Yamabiko no Tsutae 山彦之傳 The Tradition of Echoes
152 Yami no Yo ni Tsubute Utsubekarazaru Koto 闇ノ夜ニ礫ウツヘカラサル事 The Art of Not Casting Stones on a Pitch Black Night
153 Yo Maku no Koto 陽幕之事 The Art of Yo War Curtains
154 Yo Shinobi no Koto 陽忍之事 The Art of Yo Shinobi
155 Yobidashi no Jutsu 呼出シノ術 The Art of Luring Somebody Out
156 Yojihi no Koto 楊枝火之事 The Art of the Toothpick Fire
157 Yoru Iku wo Imu Koto 忌夜行事 Concerning Unlucky Days for Travelling at Night
158 Youchi no Koto 夜討之事 Concerning Night Raids
159 Yuibashi no Koto 結橋之事 The Art of the Tied Bridge
160 Yumemakura no Daiji 夢枕之大事 The Principle of the Dream Pillow
161 Zen Chu Go no Koto 前中後之事 The Art of Before, Middle and After

Opposite: Antony Cummins at the Hosa Library collection in Nagoya, Japan, where the original manuals are now kept.

頭上 注
蓬
左
文

INDEX

OTHER TITLES PUBLISHED BY THE HISTORY PRESS

In Search of the Ninja

Antony Cummins

ISBN 978-0-7524-9210-0

Lost in modern myth, false history and general misinterpretation, the ninja have been misrepresented for many years. Recently, a desire for a more historical view has emerged and here Antony Cummins fulfils that need. *In Search of the Ninja* is based upon the Historical Ninjutsu Research Team's translations of the major ninja manuals and consists of genuinely new material..

Samurai War Stories: Teachings and Tales of Samurai Warfare

Antony Cummins and Yoshie Minami

ISBN 978-0-7524-9000-7

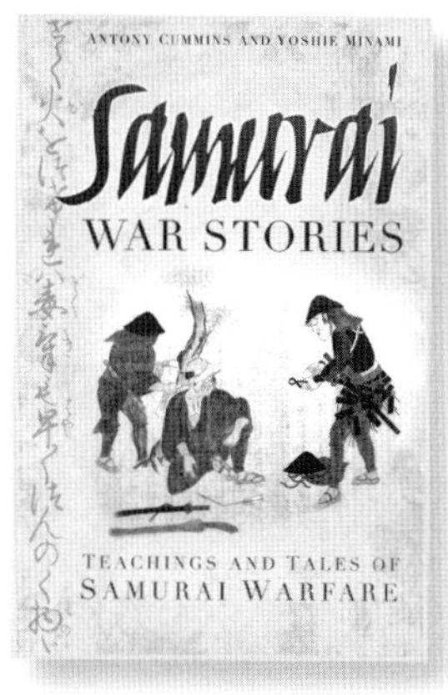

A collection of three major texts, published in English for the first time. These works include writings on three distinct military strata: the Samurai; the Ashigaru, or foot soldier; and women in war. Including guidelines, tactics, commentaries and advice written by Samurai of the period, as well as the original illustrations.

Viking Combat Techniques

Antony Cummins

ISBN 978-0-7524-8060-2

Based on a comprehensive analysis of Viking sagas and other period sources, this is the first book to present a step-by-step Viking martial system. Illustrated with over 250 photographs, this volume in effect represents the earliest fighting manual in the world.